Synthetic Materials

Second Edition

Developed by the Center for Occupational Research and Development, Inc., and sponsored by a consortium of state vocational agencies with the cooperation and support of science educators

Development: *ABC* staff, led by Project Director, Jim Cockerill

ABC Revision Committee:

Gertrude Brown, Swenson Skills Center, Philadelphia, Pennsylvania
Lisa Merritt, Crystal River High School, Crystal River, Florida
Mary Ann Figuly, De Soto High School, Kansas
Bob Farmer, Cedartown High School, Cedartown, Georgia
Jim Fromm, Kellogg High School, Coeur d'Alene, Idaho
John Reiher, Delcastle Technical High School (Science Chairperson Ret.), Wilmington, Delaware
Holly Halabicky, Genesse Independent School District, Davison, Michigan
Alvin Beadles, Thomas High School, Thomas, Oklahoma
Sue Woodfin, Benton High School, Benton, Illinois
Elizabeth Edmondson, AOP Hub, Seneca, South Carolina
Wayne Foley, Wilson School District, West Lawn, Pennsylvania
Cindy Manning, Merced High School, Merced, California

Phonetic spellings follow *The American Heritage Dictionary of the English Language, Third Edition,* Houghton Mifflin Company, 1992.

Published and distributed by:

CORD Communications, Inc.
P.O. Box 21206, Waco, Texas 76702-1206
324 Kelly Drive, Waco, Texas 76710-5709
254/776-1822 Fax 254/776-3906

Second Edition

Printed in U.S.A. May 2001

ISBN 1-57837-083-3

Preface

In a past era, a person might find or cut a piece of wood and carve a spoon out of it, or dig clay out of the ground and build a pot. In more recent times, people continued to use what was at hand to make the things they needed. Those who lived in a region rich in clay had brick factories and built their houses of brick. Even in designing something as complicated as an airplane, engineers started with the available metal—steel—and designed a shape that would fly.

Today, objects are beginning to be designed in a different way. When engineers are asked to make something, they don't necessarily start with the available material. They start with an idea for an object or a device. Then they think about how they want the device to function or behave. They design materials that will behave in the desired way. This is a very basic change in our way of making things.

The materials that we are making are different too—different from the metals and ceramics and naturally—occurring wood, cotton, and rubber that were used in the past. The new materials include polymers (many of them known to us as plastics), advanced ceramics, new metal alloys, and combinations of these materials called composites. The new materials are used in everything from the space shuttle to stereos.

At this point, you may be saying, "So what? How is this going to affect me? What difference is it going to make in my life?"

That's not only a fair question, it's a good question to ask about any new technology. In this case, the change in materials used to make things is changing what you have to know to make them. Anyone who wants to be a maker of things—a producer—will have to understand much more about the materials they are using. To design materials, you have to understand their structure—the atoms and molecules that make them up and the forces that hold them together.

On a more basic level, it's good to know about materials because, if you don't understand what your world is made of, you may begin to feel less connected to it. This sounds abstract, but it appears to be true that humans seem to enjoy having some knowledge of their surroundings.

This unit is designed to connect you with the materials revolution. If you're planning to make or buy anything in the next fifty years, this unit is for you. If you're planning to live on this planet for the next fifty years, this unit is for you. If you think that you might travel to another planet during the next fifty years, this unit is especially, most of all, for you.

Table of Contents

SYNTHETIC MATERIALS

UNIT GOALS

After you complete this unit, you will be able to—

1. Relate uses of materials to their chemical and physical properties.
2. Use the periodic table to predict how certain elements are likely to bond with others.
3. Predict the properties that will result from the use of different chemical and physical manufacturing processes.
4. Assess the effects of the chemical and physical manufacturing processes used to produce various materials.
5. Evaluate the impact of material processing and production on worker health and the environment.
6. Practice industry methods for measurement and testing of selected materials.
7. Simulate the process by which a product is designed in industry—designing for manufacturability, environmental standards, and profitability.

SUBUNIT 1

TRADITIONAL MATERIALS

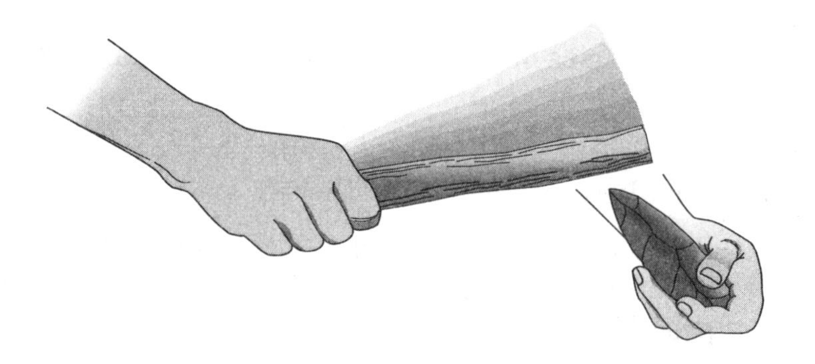

THINK ABOUT IT

- How have the uses of materials changed from prehistoric times to the present?
- How have the materials changed?
- Have these changes in materials brought about new uses?
- Are these changes good for human beings? Why or why not?

SUBUNIT OBJECTIVES

After you complete this subunit, you will be able to—

1. Select materials that might be appropriate for certain uses, based on some of the observable properties of those materials.
2. Using appropriate resources, identify materials that have specific physical and chemical properties.
3. Make sketches of known crystals and compare them to a table of crystal systems.
4. Relate different types of chemical bonding to material properties.
5. Identify five types of chemical reactions that play roles in the formation of materials.
6. Rank liquids based on the results of viscosity tests.
7. Relate the behavior of a liquid to its viscosity in a given situation.
8. Compare two methods of testing the viscosity of a liquid.
9. Grow crystals of a salt from a saturated solution.
10. Evaluate conditions that determine the size of the crystals.

PROCESS SKILLS

You will use these skills in lab—

- Position a metal object using a magnet.
- Time an event using a stopwatch.
- Measure distance using a ruler.
- Use a hand lens to magnify an object.
- Dissolve a solute in a solvent.
- Weigh a substance using a balance.

Properties and Uses of Materials

Artist and Sculptor

Jake is an artist, a sculptor who works with ***metals*** *and* ***ceramics****. His wife Lu works as a chemistry technician for an aircraft company. This Sunday afternoon they are strolling through the city's museum of fine arts. Right now they are admiring a Greek vase, made around 500 B.C. Let's listen in on their conversation.*

Jake: What a beautiful vase! I especially like these red and black Attic vases. I wonder how they got these colors.

Lu: They come from the iron in the pottery.

Jake: Both colors come from iron?

Lu: Right. The black areas had a small amount of ***alkali*** (ăl′kə-lī) *added to them to make them more glassy. In the final stage of firing, the parts of the glaze that were not glassy would oxidize, changing the form of the iron from ferrous iron to ferric iron, which created the red color.*

Jake: Wow, how did you know that, Lu?

Lu: Elementary chemistry, my dear Jake.

Jake: Wait a minute! You just read the explanation on the card next to the vase!

Lu: Well, okay, I got a little help from the card. But I do know something about the chemistry of these art objects.

Jake: Like what?

Lu: I know that in general the ***molecular*** (mə-lĕk′yə-lər) ***structure*** *of the material is related to the properties of the object and to the process of making it.*

Jake: What do you mean? Give me an example.

Lu: Okay, look at these 5th century Persian rings. The artisan chose a metal, gold, because of its strength and ***malleability*** (măl′ē-ə-bĭl′ĭ-tē)*. It's the crystalline structure of metals that allows them to be deformed without breaking them. The whole process of working the metal involves changes in the crystalline structure.*

Jake: Look, the card says that the method of manufacturing wire by drawing metal through a die hadn't been invented when this was made. The gold had to be melted and cast into an ***ingot*** (ĭng′gət). *What's an ingot?*

Lu: A mass of metal cast into a shape that makes it easier to handle, like a bar of gold.

Jake: Then it was hammered into a sheet, cut into strips, and hammered into wire.

Lu: And in the process, there probably was some ***annealing*** (ə-nēl′ ĭng)*—that's heating and cooling—to cause recrystallization and softening of the metal.*

Jake: But artists of ancient times couldn't have known the molecular structure of the materials they used. Neither do I, for that matter.

Lu: No, but you do know about the properties, or qualities, of those materials—I hear you talk about the ***density*** *of the clays that you use, the* ***ductility*** (dŭk′-tĭl′ĭ-tē) *of the metals, not to mention melting points, color, reflectance . . .*

Jake: True.

Lu: And part of the reason you don't have to know the molecular structure of the materials you use is that they're traditional materials. They've been used for thousands of years, and the craftspeople who use them know how they will change under certain conditions—heating and cooling, for example.

Jake: I guess if I were making sculpture out of the advanced ***composite*** (kəm-pŏz′ĭt) *materials that you use, I would have to know their molecular structure.*

Lu: Yes, but I hope you aren't thinking about a major change of materials. So far, I've never laid eyes on any composite materials that look as beautiful as that Greek vase.

The artist and the technician in this scenario talk about the materials used to make the art objects they like. Artists, as well as other people who make things, have to choose their materials. Sometimes the choice is dictated by what is available. For example, if you live in a forest, you are likely to build your house out of wood, but if you live by a quarry, what material might you use?

Often the choice of materials requires careful thinking by the maker. Many questions have to be answered. One of the main questions is—What is the function of the object being made? (Put another way, how is this object going to be used?) The big question may relate to some specific ones about the material such as:

- Does the material to be used need to conduct heat? Does it need to retain heat or cold?
- Does the material need to resist harsh chemicals?
- Does it need to conduct **electricity**?
- Will the material hold its shape when it is formed? How long will it last?

The maker of an object has to think about the process of making it. These questions may have to be answered:

- Will processing this material take a lot of labor or a lot of time?
- Will this material require a lot of energy to be formed into the end **product**?
- Will processing this material involve a lot of waste materials to be disposed of after the product is made?

Finally, the maker of an object has to think about the cost and availability of the material. These questions may have to be answered:

- Is this material readily available? Will it continue to be?
- How much does the material cost? How much does it cost in relation to what can be charged for the end product?

Can you think of some other questions that have to be answered? Try the following activity to help you start thinking like a designer.

Activity 1-1

- Divide the class into four groups. Assume that each group is a member of a design team. Let each group choose three objects from the list below and consider what properties are needed in the materials used to make each of them. Be as specific as you can and also specify the properties of coatings, if coatings such as paint are to be used.
  - Toy box for children under 3 years old
  - Lampshade for a conventional light bulb
  - Container for tomato plants
  - Artificial foot and ankle
  - Window covering for a window that gets full afternoon sun in a hot climate
  - A piece of luggage that has to hold food as well as clothing, and that must be lightweight enough for use when hiking.
- Share your choices with the class and, where there are differences, explain why your group listed the properties you did.
- Record your findings in your *ABC* notebook for use in the Unit Wrap-Up Activity.

In this unit, we will look at the three main types of materials—metals, ceramics and **polymers** (pŏl′ə-mərz)—as well as composite materials, made from two or more other materials together. You will learn more about these materials in Subunits 2 through 5.

In Activity 1-1 you had to consider the characteristics of the material to be used. Such characteristics are called **properties**. Materials properties can be thought about in four major classifications. They include

- **physical properties**
- **chemical properties**
- ability to be formed into desired shapes
- ability to achieve a consistent internal structure.

Any specific material will have properties belonging to all four classifications. We will look at each classification in this section.

The properties of materials can be determined through measurement or testing. In the manufacturing industries, materials measurement or testing is an important function often carried out by a technician. Materials are often tested when they come into a factory to make sure that they meet the specifications expected. Final products are tested to make sure that they will perform as their manufacturers promise.

Some tests and measurements do not affect the material or the product being tested. These are called nondestructive tests. Other tests destroy the material or part as it is being tested. These are called destructive tests. For example, to test the tread life of a tire, manufacturers have to wear down the tread until it is no longer functional. Throughout this unit, you will read more about the testing and measurement of materials and the people who carry out these tasks.

Physical Properties

Physical properties include mechanical properties, electrical properties, magnetic properties, fluid properties, thermal properties, and optical properties.

Mechanical Properties

The mechanical properties of a material are among the most important physical properties in many applications. The mechanical strength of a material influences almost every potential use. Mechanical properties include

- **tensile** (tĕn′səl) **strength**—ability of a stationary block to resist breaking by pulling forces
- **compressive strength**—ability of a stationary block to resist breaking by pushing forces
- **shear strength**—ability of a stationary block to resist breaking by forces causing an internal sliding
- **fatigue strength**—ability to resist breaking due to repeated loading and unloading

- **shock loading**—ability to resist breaking due to a sudden change in load
- **vibration characteristics**—ability to withstand vibrations
- resistance to shape changes due to elastic or plastic deformations
- resistance to formation of geometric discontinuities such as notches.

CAREER PROFILE: MATERIALS LABORATORY TECHNICIAN

Cliff L. is a laboratory assistant in a materials-testing lab. The company he works for makes many different polymer products. Cliff's job is to test samples of the finished products and verify that they have the properties they are supposed to have.

Some of the properties that Cliff tests are tensile strength, weight, density, compression strength, and hardness. Cliff runs these tests using specialized test equipment. "Some of this test equipment we made ourselves," Cliff says, referring to the other lab technician and himself. "This is a small polymer house (company) as they go. When we started up years ago, we didn't have much money to put into the lab. So we made some of the equipment ourselves—for example, that compressive strength tester." He points to a device on one of the lab tables.

"My job requires good organization and careful observation. We keep a clean lab to avoid accidents and confusion. We keep meticulous records. We check and double-check our calculations. And when there are problems, we are the ones who go to production and help troubleshoot."

Cliff used to work on the production floor making plastic parts, so he knows what can go wrong in production and how that can affect the properties of a finished product. Besides his on-the-job experience, Cliff had high school courses in physics and chemistry and a college course in organic chemistry.

Electrical Properties

Electrical properties have their greatest role in electrical equipment. They include

- **conductivity** (kŏn′dŭk-tĭv′ĭ-tē)—ability to carry an electrical current
- **resistivity** (rē′zĭs-tĭv′ĭ-tē)—ability to oppose an electrical current
- **inductive heating**—heating in a conductor due to changes in electromagnetic induction while conducting alternating electrical current
- **dielectric** (dī′ĭ-lĕk′trĭk) **heating**—heating in an insulator while opposing an alternating electrical current.

Electrical properties are closely related to magnetic properties.

Magnetic Properties

There are three kinds of magnetic materials:

- **diamagnetic** (dī′ə-măg-nĕt′ĭk)—repelled by a **magnetic field**
- **paramagnetic** (păr′ə-măg-nĕt′ĭk)—weakly attracted by a magnetic field
- **ferromagnetic** (fĕr′ō-măg-nĕt′ĭk)—strongly attracted by a magnetic field.

Ferromagnetic materials interact strongly with magnetic fields. These strong interactions make ferromagnetic materials very useful in instrumentation and with electrical power use.

Fluid Properties

Fluid properties relate to the ability of a material to flow. Some materials are useful because they are normally fluid at the operating temperature of machinery. These include lubricating oils, hydraulic fluids, and coolants. In some situations, the material used must have little resistance to flow—examples are hydraulic fluid in an automotive brake system and oil and coolant in an automotive engine. In other applications, the materials must have a resistance to flow—

examples are grease in automotive wheel bearings and the fluid in shock absorbers.

Some materials have fluid properties for only a short time, such as certain metal **alloys**. Their fluid properties allow them to be poured into molds and, when cooled, to retain the shape of the mold.

Thermal Properties

Thermal properties are related to the way a material responds to heat and cold. Thermal properties of materials include

- **thermal conductivity** (kŏn′dŭk-tĭv′ĭ-tē)—the ability to conduct heat
- **specific heat**—the amount of heat energy required to increase the temperature of one gram of the material by one Celsius degree
- **thermal expansion**—the amount of change in volume that results from a change in temperature
- **melting point**—the temperature at which the material changes from a solid to a liquid
- **boiling point**—the temperature at which the vapor pressure of the material equals one atmosphere of pressure (101,325 pascals or 14.7 psi).

Thermal properties must be considered carefully if a material will be used under a variety of temperature conditions.

Refractory (rĭ-frăk′tə-rē) materials are materials that are especially stable at elevated temperatures and do not melt or decompose until the temperature becomes extremely high—for example, 3000°F (1649°C). Can you think of some materials that might be in this category? What might be some uses for refractory materials?

Optical Properties

Optical properties affect the way materials absorb, reflect, or transmit light. For example, the optical property of reflectance is important in applications such as mirrors, fiber-optic cables, and retroreflectors.

Activity 1-2

- Divide the class into teams of two. Assign each team a physical property from the list below. (In a class of 28 or more, at least two teams should work on each property.)
  - Tensile strength
  - Fatigue strength
  - Conductivity
  - Inductive heating
  - Ferromagnetism
  - Specific heat
  - Melting point
- Design a way to test materials for the given property. You do not need to carry out the test, but you should have specific ideas about what equipment would be required and what procedures would be followed. Develop the equipment and procedures list in your *ABC* notebook and for class presentation.
- Compare your team's solution with that of the other team. Select the best solution (or combine your efforts to develop a better solution) and present it to the class.
- Record your findings in your *ABC* notebook for use in the Unit Wrap-Up Activity.

Chemical Properties

The chemical properties of a material relate to its ability to enter into or resist a chemical change. When a substance undergoes a chemical change, its atoms are rearranged and/or added to new atoms to form a different compound. Chemical properties include **reactivity**, stability, **corrosion** resistance, **flammability**, **reducibility**, and **oxidizability** (ŏk′sĭ-dīz′ə-bĭl′ĭ-tē), to name a few. Each of these properties implies a change (or resistance to change) of a material into another material. Many chemical properties cannot be measured without subjecting them to conditions that are potentially destructive.

Reactivity and Stability

Reactivity and stability represent opposite extremes of the same property—the tendency to interact with other materials. A reactive compound has low stability, and a stable compound has low reactivity. Some metals, such as lithium, sodium, and potassium, are very reactive. They react quickly and explosively with water and are easily oxidized. At the other extreme, helium, neon, and argon are very stable. They do not react under normal conditions and are called **inert gases**. (They are also called noble gases.)

Oxidizability and Reducibility

Oxidizability and reducibility represent opposite extremes of a specific type of reactivity—reactivity to **oxidation-reduction** (ŏk′sĭ-dā′shən rĭ-dŭk′shən) or **redox** (rē′dŏks′) reactions. If a material has high oxidizability, it can easily lose one or more electrons during a reaction. Then we would say that the material can be oxidized easily. If a material has a high reducibility, it can easily gain one or more electrons during a reaction. In other words, it can be reduced easily.

Flammability

Burning or combustion is one example of an oxidation-reduction reaction. In a combustion reaction, oxygen is reduced and the fuel is oxidized. A flammable material is one that can be quickly oxidized by oxygen.

Corrosion Resistance

Corrosion resistance represents a material's ability to resist **erosion** by oxidation or by **acids**. Corrosion resistance in a material indicates a specific kind of chemical stability.

Activity 1-3

- You are probably already familiar with many metals, ceramics, and polymers. In your *ABC* notebook, list some familiar metals such as copper, zinc, iron, gold, and so on. List some familiar ceramics such as porcelain, stoneware, brick, and glass. List some familiar polymers such as rubber, wood, cotton, polystyrene (pŏl′ē-stī′rēn), and polyurethane (pŏl′ē-yŏŏr′ə-thān′).
- Make a table of these familiar metals, ceramics, and polymers.
- Beside each material listed, write one or more uses for that material. Beside each use, indicate the property or properties that make the use possible.

  For example, copper is used in electrical wiring. Two properties of copper that make possible its use for wiring are electrical conductivity and its **ductility** (dŭk′-tĭl′ĭ-tē).
- Compile the tables made by everyone in the class.
- As a class, try to decide what properties the materials in each group—metals, ceramics, and polymers—have in common.
- As a class, try to decide which properties are chemical properties and which are physical properties.
- Record your findings in your *ABC* notebook for use in the Unit Wrap-Up Activity.

Ability to Be Formed into Desired Shapes

The use of a material depends upon a suitable method for changing the solid material into a predetermined shape with accurate dimensions. Four methods used to form a material into a desired shape include

- **machining**
- **welding**
- **casting**
- using the plastic properties of the material.

Objects formed by these methods are illustrated in Figure 1-1.

Each of these refers to a different method of shaping material. These are not the only methods of shaping available; they are simply among the most commonly used.

Figure 1-1 Ability to be formed into desired shapes

Machining

The shaping method most familiar to many people involves removing chips of material by mechanical means. This includes milling, drilling, and cutting on a lathe. But these are not the only machining methods. Machining involves the shaping of a material by removing the undesired portions. Other methods of machining include burning, chemical dissolving or etching, electric spark eroding, and electrolytically dissolving material.

Welding

Welding involves forming a single piece of material from two separate pieces. Welding requires the material to melt and to mix with the other piece of material or the filler material from the welding rod. Welding requires properties that are similar to the properties required for casting and using the plastic properties of the material.

Casting

Casting involves the solidification of a liquid material into a predetermined shape controlled by a form or a mold. One notable exception to the heating requirement is concrete. Liquid concrete is mixed, poured into the forms, and then sets to a solid (without external heating) as chemical reactions take place. Casting often involves reheating a metal or other material. The reheating process often affects the crystalline structure of the material and changes its characteristics.

Using the Plastic Properties of the Material

The plastic properties of a material are its ability to be shaped by forces without breaking the material or damaging its internal structure. (Plastic, in this case, does not mean made of plastic. It relates to the ability of the material to flow.) Shaping methods that use the plastic properties of materials include bending, spreading, stretching, and extruding.

CAREER PROFILE: CERTIFIED PROSTHETIST/ ORTHOTIST

Bonn B. is a certified prosthetist (prŏs′thĭ-tĭst)/orthotist (ôr-thŏt′ĭst). His job is to make artificial limbs for people who have, for one reason or another, had an amputation (the cutting off of a limb). People lose limbs for various reasons, including diseases such as diabetes that cause poor circulation, trauma, **cancer** (kăn′sər), and **congenital** (kən-jĕn′ĭ-tl) (from birth) deformity.

A prosthesis (prŏs-thē′sĭs) is an artificial device to replace a missing part of the body. Limb prostheses are usually made of two or three parts: a socket that fits over the residual limb (the part of the limb that remains after the amputation), a structural part that bears weight and extends the limb, and an artificial foot or hand that fits at the end of the prosthesis. Bonn's job is to evaluate the patient's needs, fit and make the socket, select the components, and assemble the prosthesis.

"During the initial evaluation, I talk to the patients about their lifestyle—hobbies, sports, profession, and so on. A guy who likes to swim and ride a bicycle needs something different from a guy who spends most of his time in a car or at his desk. When I select the components of the prosthesis, I have to consider lifestyle as well as weight, skin sensitivity, the shape of the patient's residual limb, and so on."

Bonn explains that the socket of a prosthesis is like a shoe for the residual limb, and that has to fit perfectly. Fitting and making the socket is "very hands-on," he says, and it's obvious that he especially enjoys this part of the job. He takes a plaster-of-paris cast of the residual limb, makes a positive mold of that out of plaster, smoothes and shapes the positive mold, and makes a check socket out of a clear plastic resin. The patient then returns to see if the check socket fits properly. If the socket fits, another positive mold is made, and the prosthetic socket is made from that.

The prosthetist's job requires knowledge and skills in several different areas. Bonn earned a bachelor's degree in orthotics-prosthetics studying human anatomy, **physiology** (fĭz′ē-ŏl′ə-jē), and pathology. "In addition to my knowledge in science, I have to be a craftsman to work casting, carving, and smoothing the test socket. But it helps to understand materials on a molecular level—the thermoplastics, for example. It's important to get all of the plastic heated evenly. If you don't, some chains of molecules are moving too fast while others are still bound together. That puts stress on the plastic and makes it brittle."

Another part of the job that involves a knowledge of materials is selection of the structural components that are attached to the socket. In a below-the-knee prosthesis, for example, the structural component would be the part between the socket and an artificial foot. "The structural component might be made of wood, aluminum, carbon, graphite, titanium, or other material," says Bonn. "We don't make structural components or artificial feet here. We select them.

"Artificial feet come in many types—wood, plastic, and so on. The patients' needs dictate our selection. That's true with every component." Bonn goes on, "For example, we typically use a polyethylene called 'pelite' to line the socket. But if a patient is prone to skin breakdown—the formation of abrasions and blisters—as many people with diabetes are, we use a silicone liner instead. The silicone liner absorbs the shear forces exerted on the skin and helps to prevent skin breakdown."

Bonn describes his job as "a lot of fun and pretty satisfying. People come in on crutches or in wheelchairs; they leave walking."

Ability to Achieve a Consistent Internal Structure

The ability to achieve a consistent internal structure depends on the atomic or molecular structure of the material, its **crystal** structure, and the size of its crystals. For a pure, solid substance, the most common form is **crystalline** (krĭs′tə-lĭn). Crystalline means that the **atoms** or **molecules** (mŏl′ĭ-kyo͞olz′) are arranged in fixed positions with specific distances and angles between neighboring atoms or molecules. The properties of most solid materials are the result of their crystalline structure. Some of the materials that you will study in this unit, such as metal alloys and ceramics, have a crystalline structure.

Crystal Structure

Activity 1-4

- Spread some NaCl (sodium chloride) crystals on a sheet of white paper. Examine these crystals of salt using a hand magnifying glass (or stereomicroscope if one is available). Draw a sketch of the crystals in your *ABC* notebook and label the sketch "NaCl crystals."
- Spread some sucrose (table sugar) crystals on a sheet of white paper. Examine these crystals using a hand magnifying glass (or stereomicroscope if one is available). Draw a sketch of the crystals in your *ABC* notebook and label the sketch "sucrose crystals."
- In your *ABC* notebook, write a brief description of the differences between the salt crystals and the sugar crystals. If you knew an unknown substance was either salt or sugar, could you identify it based on the shape of the crystals?

There are seven crystal groups or systems based on the shape of the crystal. Table 1-1 summarizes these crystal systems. The geometry of the crystal determines its system. To which crystal system does table salt belong?

Table 1-1. Seven Crystal Systems

Shape	Crystal System	Angles Between the Crystal Axes	Lengths of the Crystal Axes
	Cubic	All are 90°	All are equal
	Tetragonal	All are 90°	2 are equal 1 is unequal
	Hexagonal	1 is 90° 3 are 60°	3 are equal 1 is unequal
	Rhombohedral	All are not 90°	All are equal
	Orthorhombic	All are 90°	All are unequal
	Monoclinic	2 are 90° 1 is not 90°	All are unequal
	Triclinic	All are not 90°	All are unequal

Table salt crystals belong to the cubic system. However, when you looked at the magnified salt crystals, you probably saw rounded edges on the cubic structures rather than the sharp edges you would expect on a cube. The instability of the atoms or molecules along the edges and at the corners of the crystal causes this rounding. Consider the atom or molecule at the corner of a perfect cubic crystal as shown in Figure 1-2. The crystal in this position has three surfaces exposed. As a result, free atoms (atoms still in solution) or solvent molecules can collide with the corner atom from three directions. Any collision with enough **kinetic** (kĭ-nĕt′ĭk) **energy** can knock the corner atom out of its position and remove it from the crystal, that is, dissolve the atom or molecule.

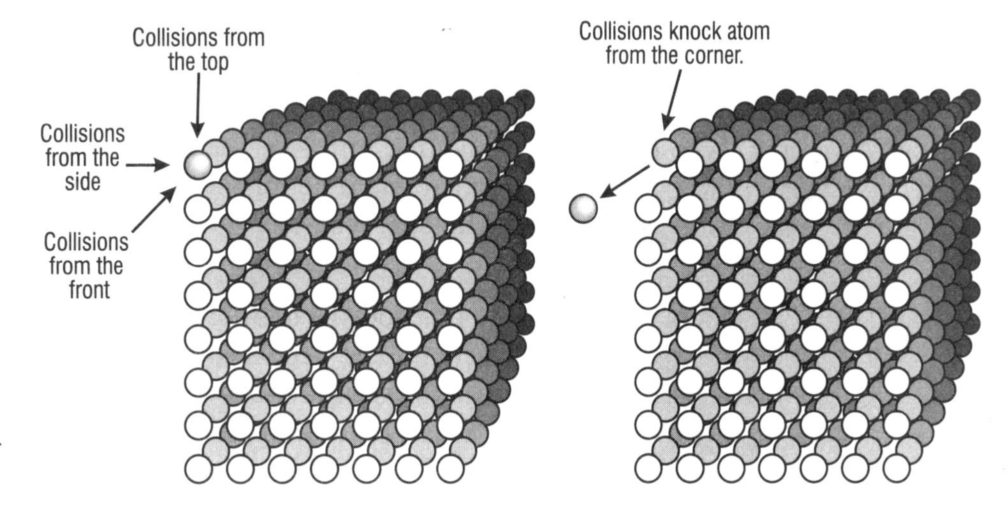

Figure 1-2 Instability of corner position of a cubic crystal

The atoms along the edges of the cube have two surfaces subject to collisions. The atoms in these positions have more stability than the atoms in the corner position. However, they have less stability than the atoms on the surface of the cube, which have only one surface exposed to collisions. Since the corner and edge positions have little stability, the cubic crystals (and other crystal classes) have imperfect shapes. The corners and edges are rounded because collisions have knocked loose the atoms or molecules that occupied these positions.

Crystal Formation

Most materials exist as solids in some crystalline form. The way crystals form has an important effect on how materials behave. Many materials and products are made by controlling the formation of crystals. For example, steel exists in three different forms, with each form having a different crystal structure, brought about by different cooling processes.

Crystals form under two conditions. They grow when a melted compound cools. They also grow from **saturated solutions**. The more slowly the crystals form, the larger they become.

For crystal growth to begin, a **nucleus** (no͞o′klē-əs) must form first. (In this case, nucleus does not refer to the nucleus of an atom.) This initial nucleus is a small crystal. When crystals are grown from a solution, a small "seed crystal" is often added to the solution to act as a nucleus for crystal growth.

As a melted compound or "melt" cools, the atoms or molecules lose thermal energy and move more slowly. When they move more slowly, collisions between atoms or molecules can cause them to stick together. As more atoms or molecules collide and stick, the initial nucleus for a crystal forms. When the nucleus forms, atoms or molecules that collide with the surfaces of the nucleus add to the growing crystal.

Different methods of heat treating and **quenching** have different effects on the properties of material. Slow cooling allows crystals to become large. Quick cooling results in the formation of small crystals. Differently sized crystals give the material different properties.

Activity 1-5

- Follow the instructions given in the handout provided by your teacher and grow your own crystals. After 10 days bring your crystals to class and compare your results.
- Examine your crystals under magnification.
- Identify whose crystals were largest and whose were smallest and try to determine what each student did differently to produce different-sized crystals.
- You may eat your results.

Elements—The Basic Materials

Where Does Everything Come From?

Jake and Lu, the artist and chemistry technician you met in the first scenario, have a three-year-old son named Ramsey. At this age, Ramsey is especially curious about where things come from. Today he's hanging around in Jake's workshop, asking questions, as usual.

Ramsey (picking up a piece of scrap metal from the floor): What's this, Daddy?

Jake: Oh, that's a piece of steel.

Ramsey (looking at the scrap metal): Where does steel come from?

Jake: It's a metal—well, it comes from iron and ...

Ramsey: And what?

Jake: Iron and another material called carbon.

Ramsey: Where do iron and carbon come from?

Jake: They come from the earth—out of the ground; they're called ***elements****. Elements are the basic materials of which everything is made.*

Ramsey: Am I made out of elements?

Jake: Yeah, but you have many more elements than steel.

Ramsey (picking up a piece of a clay pot): Where does this come from?

Jake: That's made of clay, Ramsey. Clay comes out of the ground.

Ramsey: Is clay an element, too?

Jake: No, but it's made of elements—iron, aluminum, silicon, and oxygen.

Lu (appearing in the doorway): Gee, Jake, do you think he can understand all that?

Jake: No, but right now he wants names for things, so I might as well tell him the right names.

Lu: I guess you're right. Listen, I just warmed up the spaghetti you made last night, Jake. How about taking a break for lunch?

Ramsey: Where does spaghetti come from?

Lu (walking into the kitchen with Ramsey): Let's see. . .the noodles come from wheat, and the wheat grows in the ground. The sauce is made from tomatoes and . . .

As Jake told Ramsey in the preceding scene, elements are the basic materials from which everything is made. Ninety-one elements occur naturally on Earth. However, most materials are not made up of a single element. They are made of two or more elements combined to form **compounds**.

What Causes Elements to Interact as They Do?

You may know from previous studies that an atom is the smallest particle of an element that has all the properties of that element. An atom consists of **electrons** and a **nucleus** (nōō′klē-əs) (made of **protons** and **neutrons** [nōō′trŏnz′]). Electrons have a negative charge. Protons carry a positive electrical charge, and neutrons do not have a charge.

When you are trying to figure out how an element is going to behave, you need to know more than how many protons, neutrons, and electrons it has. You need to know something about how the electrons are arranged. The regions of space around the nucleus in which electrons can locate are called **orbitals**. Some orbitals are relatively close to the nucleus; these are referred to as lower energy levels because it requires less energy for electrons to occupy them than it does to occupy orbitals that are farther from the nucleus. In some atoms, only one energy level—the one closest to the nucleus—is occupied. In other atoms, more than one energy level or orbital is occupied.

Electrons usually occupy the orbitals that require the least energy. Electrons in the outermost orbitals tend to have more energy and to be more available to interact with other atoms. That is, they can attract and form bonds with other atoms.

Materials—whether they are elements or compounds—behave the way they do in large part because of the atomic structure of the elements they contain. So far in this unit, we have used the word properties to talk about the way materials behave. Properties such as electrical conductivity or flammability result from the way atoms in a material interact. Atoms interact based on their atomic properties; that is, the number of protons and electrons they have and the orbitals in which the electrons are located. This is true whether the material is a basic element or a very complex compound containing many elements.

Chemists and other people who deal with materials have to know quite a bit about what is going on with materials at the atomic level. How do they organize and keep up with ninety-one elements and an endless number of compounds? One tool on which they rely is the **periodic table**.

What Can You Find Out from the Periodic Table?

The periodic table is an arrangement of all known elements. The periodic table was first proposed by a Russian chemist, Dmitri Mendeleev (də-mē′trē mĕn′də-lā′əf), in the 1860s, Mendeleev arranged the known elements based on their atomic masses. What is the **atomic mass**? The atomic mass of an element in atomic mass units (AMU) is the sum of the number of protons and the number of neutrons in the nucleus of an atom of the element.

Mendeleev organized the first two rows of elements into seven elements each and the second two rows into seventeen elements each. This arrangement gave a vertical grouping of elements with similar properties. However, there were some blank spaces in the table. Mendeleev was able to predict that elements with certain properties would eventually be discovered and would fit into the blank spaces, according to their atomic masses.

Mendeleev's arrangement had an inconsistency, however. Iodine and tellurium (tĕ-lo͝or′ē-əm) seemed to be in the wrong columns based on their properties. The modern periodic table arranges the elements on the basis of **atomic number** rather than atomic mass. With this arrangement, iodine and tellurium fit in the correct columns based on their properties. What is the atomic number? The atomic number of an element is the number of protons in the nucleus of an atom of the element.

Periods and Groups

The modern periodic table arranged by atomic numbers is shown in Figure 1-3.

The horizontal rows of elements are called periods. For example, hydrogen and helium make up the first period and lithium, beryllium, boron, carbon, nitrogen, oxygen, fluorine, and neon make up the second period. As you go from the left side of a period to the right side, the atomic number increases by one for each element.

The vertical columns of elements make up families or groups. The groups are numbered with numbers 1 through 18. Group 1 consists of hydrogen, lithium, sodium, potassium, rubidium (ro͞o-bĭd′ē-əm), cesium, and francium (frăn′sē-əm). The families are referred to by names such as inert gases for group 18 and halogens for group 17. Elements in the same group have a similar number of electrons in their outermost orbitals, and these elements have similar properties.

The modern periodic table contains 112 elements. Ninety-one of these are naturally occurring on Earth; the others are made in the laboratory.

Besides the division periods and groups, the elements in the periodic table are sometimes roughly divided into two major classifications: metals and **nonmetals**. Of the 112 elements, 90 are metals and 22 are nonmetals.

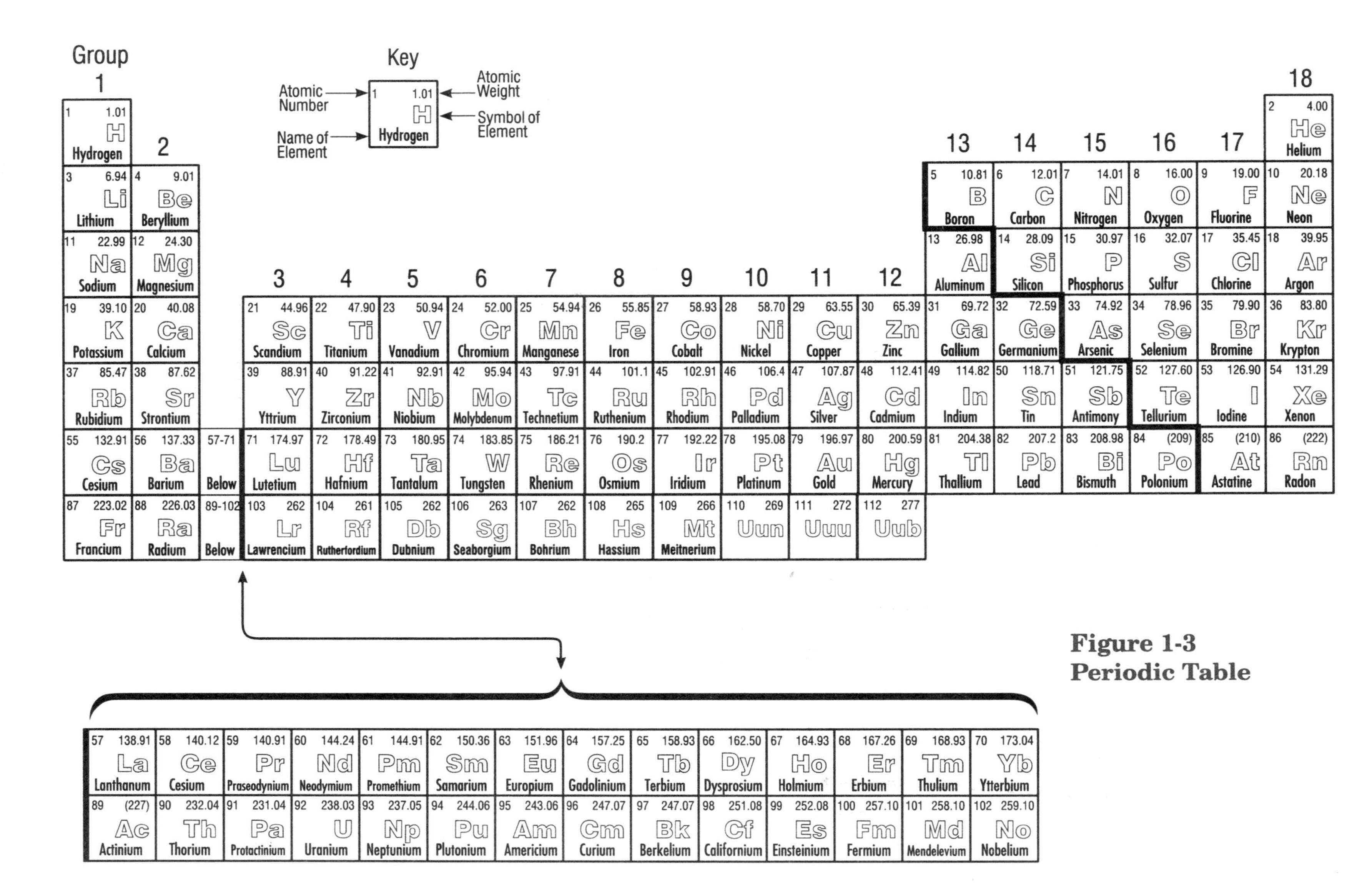

Figure 1-3
Periodic Table

The line on the right side of the table that begins next to boron (B) and stair-steps down to the right separates the metallic elements from the nonmetallic elements (except hydrogen, which is a nonmetal but is placed in Group 1 with lithium, sodium, potassium, and so forth).

Atomic Properties

Metallic atoms have certain properties in common. When one is bonded to another atom, it does not show a strong attraction for the shared electrons—a tendency known as low **electronegativity** (ĭ-lĕk′trō-nĕg′ə-tĭv′ĭtē). The outer electron is easily removed, creating a positive ion; that tendency is known as low **ionization** (ī′ə-nĭ-zā′shən) **potential**. The outer electrons are not strongly attracted to the nucleus—that is, metallic atoms have a large **atomic diameter** since the outer electron has a larger space for its orbit.

What is electronegativity? This property of atoms describes the attraction of an atom for the shared electrons when it is bonded to other atoms. The electronegativity increases as the elements proceed through a period—that is, as the atomic number increases. The electronegativity decreases as the elements proceed through a group. This means that the most electronegative elements are found at the upper right corner of the periodic table and the least electronegative elements are found at the lower left corner of the table. Fluorine is the most electronegative element.

What is ionization potential? This property of atoms measures the energy needed to remove an electron, thus creating a positive ion. The ionization potential of an element correlates closely with electronegativity, and the same trends follow in the periodic table. That is, the ionization potential increases as elements proceed through a period and decrease as elements proceed through a group.

What is atomic diameter? This property describes the size of the atom as determined by the volume occupied by the outer electrons. Atomic diameter is inversely related to electronegativity since the greater the attraction of an atom for an electron, the closer the electron will be to that atom and the smaller volume it will occupy. The atomic diameter decreases as the elements proceed through a period.

However, it increases as the elements proceed through a group because there are more total electrons.

Activity 1-6

- Form the class into two teams—metals and nonmetals. After deciding which team begins, the teacher asks a question that can be answered using the periodic table. If the team answers the question correctly, it receives 6 points. If the first team misses the question, it goes to the other team where a correct answer gets them 5 points. If the second team misses the question, it goes back to the first team for 4 points for a correct answer. The question alternates back and forth until one team correctly answers it or for 3 rounds, with the point value dropping 1 point for each incorrect answer.
- The second question initially goes to the second team, again with a 6-point initial value. The team that gets the most points wins the competition and advances to become noble gases.

Chemical Bonding

The atomic structure of elements is really only the beginning of the story of why materials have certain physical and chemical properties. Materials form through the bonding of atoms to other atoms. To understand materials, you have to understand the bonding characteristics of different types of materials.

Bonding

Bonding is the force that holds atoms together in molecules. Atoms consist of a positively charged nucleus surrounded by negatively charged electrons. Bonding between atoms arises from the attraction of the positively charged nuclei for the negatively charged electrons. Three types of bonding that are important in holding atoms together in molecules include **ionic bonding**, **covalent bonding**, and **metallic bonding**.

Ionic Bonding

Ionic bonding is the bonding between **ions** (positively or negatively charged atoms). It is the mutual attraction of a positive **cation** (kăt′ī′ən) (positive ion) and a negative **anion** (ăn′ī′ən) (negative ion). This kind of bonding is prevalent in salts such as sodium chloride. Many ceramic materials are held together by ionic bonding. Figure 1-4 shows ionic bonding in a sodium chloride crystal.

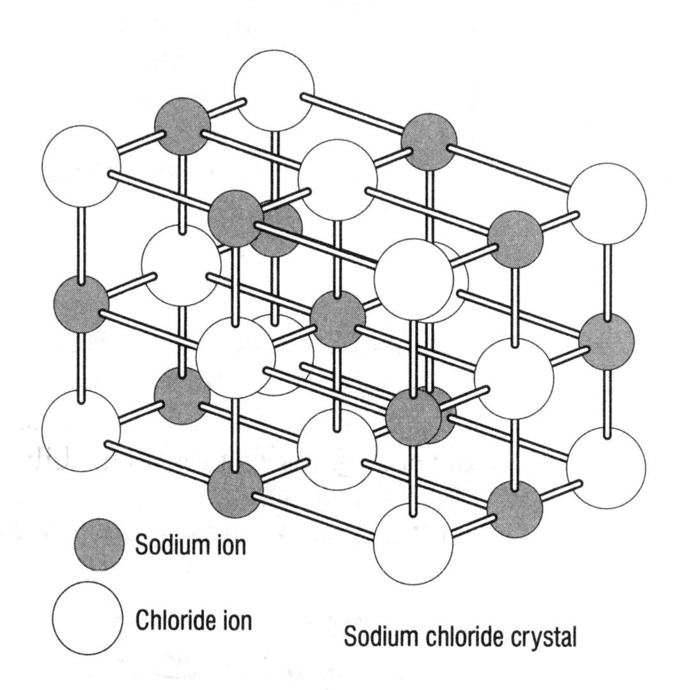

Figure 1-4 Ionic bonding in sodium chloride crystals

Covalent Bonding

When two atoms share electrons about equally between them, this is called covalent bonding. The shared electrons are attracted to the positive charge of the two nuclei that are sharing them. Covalent bonding is important in polymers and in some ceramic materials. Figure 1-5 shows covalent bonding between two hydrogen atoms.

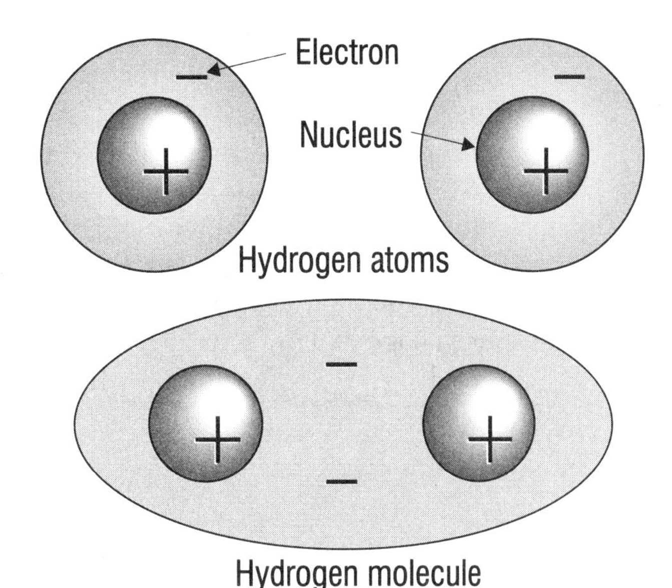

Figure 1-5 Covalent bonding in a hydrogen molecule

Metallic Bonding

In metallic bonding (Figure 1-6), the electrons are free to move around but are attracted to the positive charges of the nuclei of metal ions. This freedom of the electrons to move around gives metals their ability to conduct **electrical current** as well as other useful properties. Metallic bonding will be discussed in more detail in Subunit 2.

Figure 1-6 Metallic bonding

Weaker Attractions

In addition to these three kinds of bonding, two weaker attractions hold molecules together. The first of these is hydrogen bonding. When one part of the molecule has a partially positive hydrogen atom and another part of the molecule has a partially negative oxygen or other electronegative atom, a hydrogen bond can form between the two parts of two molecules. The hydrogen bond is the sharing of the partially positive hydrogen atom between two partially negative oxygen atoms on different molecules. Figure 1-7 shows hydrogen bonding between water molecules.

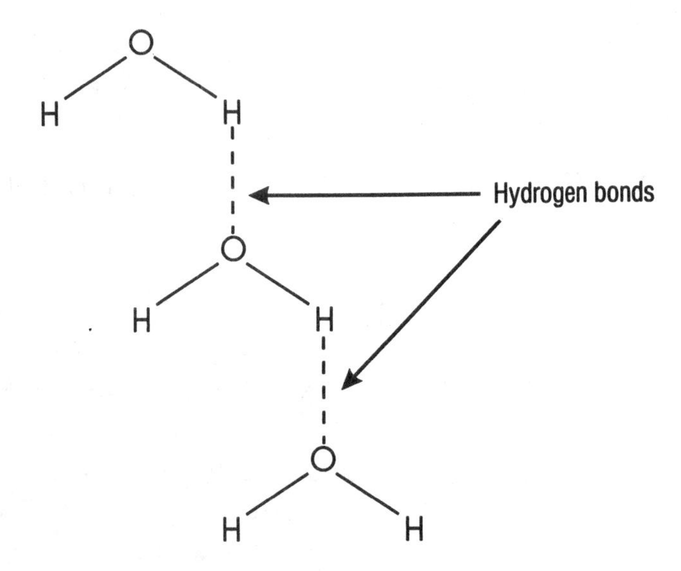

Figure 1-7 Hydrogen bonding between water molecules

The other kind of weak attraction is called **van der Waals attraction**. This comes about when the electrons of one molecule polarize the electrons of another molecule (repel the electrons and push them away). By repelling the electrons, the electrons set up a dipole, a separation of charges. The electrons are attracted to the positively charged portion of the dipole. Figure 1-8 shows van der Waals attractions between two spheres. Van der Waals attractions, the weakest type of attractions, occur in many different materials.

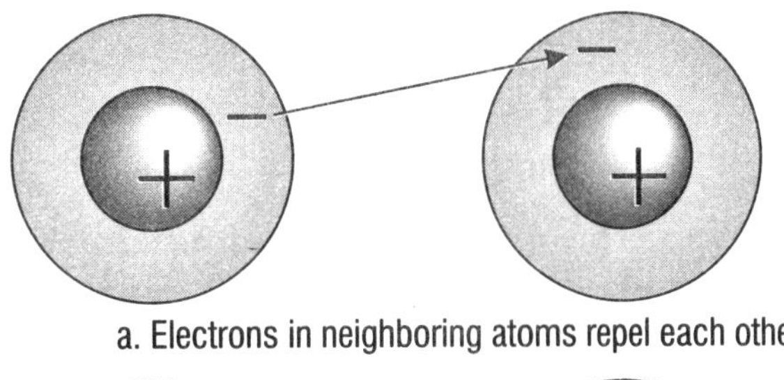

a. Electrons in neighboring atoms repel each other.

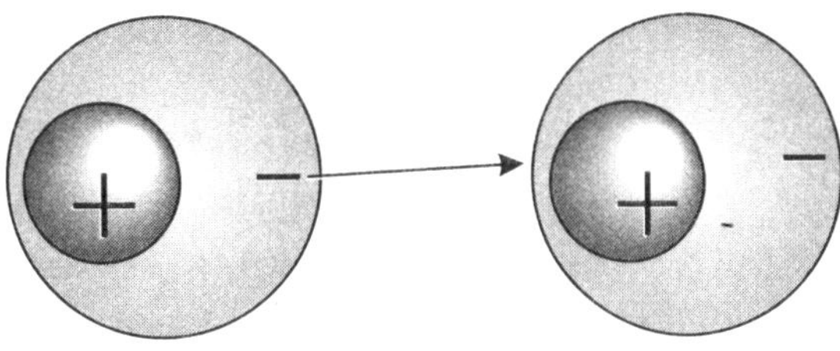

b. Electron is attracted to nucleus of neighboring atom.

Figure 1-8 Van der Waals attraction between two atoms

Chemical Reactions

In the next four subunits, you will read about how materials are made and how they interact with other materials. To understand these processes, you need to know something about **chemical reactions**. This section of the subunit will explain what a chemical reaction is and will describe several types of chemical reactions.

A chemical reaction involves changing one or more substances called **reactants** (rē-ăk′tənts) into one or more substances called **products**. In the usual method of writing a chemical reaction, reactants and products are indicated by their chemical formulas. The formulas for the reactants appear on the left side, and the formulas for the products appear on the right side. An arrow that points from left to right connects the two sides. The arrow means that the reaction of the reactants on the left "yields" the products on the right. This notation is called a **chemical equation**, and a general form of a chemical equation is shown in Equation 1-1 where A and B are reactants and C and D are products.

$$A + B \rightarrow C + D$$

Equation 1-1

Classification of Chemical Reactions

Most chemical reactions can be classified in one of the following ways:

- **single displacement**
- **double displacement**
- **decomposition**
- **combination**

In the following equations, the "aq" notation indicates the compound is in an aqueous or water solution. "g" indicates gas; "l" indicates a liquid. (An aqueous solution differs from a liquid. A liquid is a pure substance, but an aqueous solution is a mixture.) The "s" notation indicates the compound is in a solid form.

Single Displacement

A single displacement reaction involves one element replacing another in a compound, producing a new compound and a different element. Equation 1-2 shows the general form of a single displacement reaction. Equation 1-3 gives an example.

$$\text{element} + \text{compound} \rightarrow \text{compound} + \text{element}$$

Equation 1-2

$$Cl_2\ (g) + 2KBr\ (aq) \rightarrow 2\ KCl\ (aq) + Br_2(l)$$

Equation 1-3

Double Displacement

The characteristic of a double-displacement reaction involves the exchange of positive and negative ions between two compounds. Usually one of the product compounds does not dissolve so the precipitate is gaseous, indicating that a reaction has occurred. Equation 1-4 shows the general form of a double-displacement reaction.

$$\begin{matrix}\text{compound} \\ \text{AB (aq)}\end{matrix} + \begin{matrix}\text{compound} \\ \text{CD (aq)}\end{matrix} \rightarrow \begin{matrix}\text{compound} \\ \text{AD (s, l, g)}\end{matrix} + \begin{matrix}\text{compound} \\ \text{CB (aq)}\end{matrix}$$

Equation 1-4

One common type of double-displacement reaction is the **acid-base neutralization** (no͞o′trə-lĭ-zā′shən) reaction. The products of an acid-base neutralization reaction are always a salt and water. A salt consists of a cation (positive ion) from a **base** and an anion (negative ion) from an acid. Equation 1-5 shows the reaction between hydrochloric acid (HCl) and sodium hydroxide (NaOH).

$$HCl\ (aq) + NaOH\ (aq) \rightarrow H_2O\ (l) + NaCl\ (aq)$$

Equation 1-5

Decomposition

The characteristic of a decomposition reaction is a single compound reacting to form two or more simpler compounds. Equation 1-6 shows the general form of the decomposition reaction.

$$\text{compound} \rightarrow \text{two or more elements or compounds}$$

Equation 1-6

The electrolysis of water shown in Equation 1-7 is an example of a decomposition reaction. The notation "elec" above the arrow indicates that electricity is supplied to the reaction to cause the decomposition.

$$2\ H_2O\ (l) \xrightarrow{\text{elec}} 2H_2\ (g) + O_2\ (g)$$

Equation 1-7

Combination

In a combination reaction, two or more elements or compounds combine to form a new compound. Equation 1-8 shows the general form of this reaction.

$$\begin{matrix}\text{element or} \\ \text{compound}\end{matrix} + \begin{matrix}\text{element or} \\ \text{compound}\end{matrix} \rightarrow \text{compound}$$

Equation 1-8

Two synthesis reactions are important in polymers. The first is called an addition reaction. With the addition reaction, an initiator reactant (a molecule with a reactive unpaired electron) and a **monomer** (mŏn′ə-mər) react in the first step shown in Equation 1-9, forming an activated polymer. ("Mono" means one and "poly" means many.) The activated polymer adds to another monomer as shown in Equation 1-10.

Equation 1-10 can continue until the activated polymer is deactivated by another activated polymer or initiator molecule as shown in Equation 1-11. The activated polymer goes through Equation 1-10 many times, producing a large polymer molecule before it is deactivated.

monomer + initiator → activated polymer

Equation 1-9

monomer + activated polymer → activated polymer

Equation 1-10

activated polymer + activated polymer or initiator → polymer

Equation 1-11

The other type of synthesis reaction important in polymers is the **condensation reaction**. This reaction involves the joining of two molecules with the elimination of water. The reactant molecules are usually carboxylic acid, alcohol, or amine molecules. The first step of this reaction is shown in Equation 1-12. (Note that "mono" means one and "di" means two. Therefore, monomer means one "mer" and dimer means two "mers.") In subsequent steps, monomers and dimers add to the growing polymer.

monomer + monomer → dimer + H_2O

Equation 1-12

Activity 1-7

- Form the class into groups of three.
- Your teacher will give each group a piece of paper with an equation written on it. Do not show this equation to anyone outside your group.
- In each group, devise a way to present the equation in terms other than a chemical formula. You may present a skit in which students represent elements or compounds. You may use objects or pictures or some other form of presentation.
- Present the equation to the class.
- As a class, decide what type of equation was represented: single displacement, double displacement, decomposition, or combination.

Energy Minimization

Chemical reactions tend always to go in the direction of lower energy. Any system in nature changes to minimize the energy in the system. When a system has the minimum energy, it is the most stable, that is, the least likely to change. When a chemical reaction occurs, the products of the reaction have less energy than the reactants under the conditions of the reaction. Most reactions are reversible, that is, under the right conditions, the products can change back into the reactants. (The reaction will go to the left side of the equation instead of the right side.) Many reactions develop an **equilibrium** between the reactants and products of the reaction. An example of this is the ionization of water shown in Equation 1-13.

$$H_2O \rightarrow H^+ + OH^-$$

Equation 1-13

If an equilibrium develops, it can be disturbed by changing the conditions of the reaction—pressure, temperature, pH, or concentration of one or more reactant or product. When the equilibrium is disturbed, it will re-establish at new concentrations of reactant and product.

Looking Back

The properties of materials are its characteristics. Materials have physical and chemical properties. The properties of materials are related to the atomic structure of those materials and the kinds of molecules they form through chemical bonding.

The three main types of chemical bonding are ionic, covalent, and metallic. Two kinds of weaker attractive forces—hydrogen bonding and van der Waals bonding—also play a role in the way atoms and molecules are held together.

Materials are found or made through chemical reactions. At least five kinds of reactions are important in the making of materials: single displacement, double displacement, decomposition, synthesis, and redox reactions.

Further Discussion

- Select a group of the periodic table of elements. Investigate these elements and try to determine what they have in common.

Activities by Occupational Area

General

Lost-Wax Technique

- Visit a jewelry manufacturer or craftsperson or sculptor who does his or her own casting. Find out about the "lost-wax" casting technique. Make a series of diagrams in your *ABC* notebook to show how this technique works.

Agriscience

Corrosion Resistance in Farm Implements

- Contact a farm implement dealer and find out what kind of metal is used in many of the farm implements—hand tools as well as tractor attachments—that come in direct contact with the earth. How must they be maintained in order not to become rusted or corroded? What are the other properties of the materials used to make these implements? Write a report in your *ABC* notebook.

Health Occupations

Cast Materials

- Discuss with an orthopedic doctor or nurse the materials used for the casts on broken and sprained limbs. Find out the following: What materials are used to make these casts? How do the materials interact with the skin? How well do they resist water and moisture? How strong are they?

- On the basis of your discussion, write in your *ABC* notebook a brief list of things you think people need to know about the care of casts if they have to wear them.

Family and Consumer Science

Cookware Comparison

- Visit a cookware store or hardware store that carries several different brands of cookware (pots, pans, and so on). Study the product information available and ask questions of the sales personnel (be sure to visit when there are not many customers in the store) to find out about the properties of each line of cookware. Make a chart in your *ABC* notebook to show each brand, the material of which it is made, the properties of each, the aesthetic qualities, and the cost. On the basis of the information in the chart, decide which brand you would buy if you needed new cookware.

Industrial Technology

Plumbing Materials

- Contact a plumber and ask him or her about the materials normally used for household plumbing. Find out the following: What materials are used for water pipes? Are the same materials used for wastewater as for incoming water? What should you not put down your sink or toilet because it might cause wear or corrosion of your pipes? How have the materials used in home plumbing changed over the years? How can you tell what plumbing materials are used in your own home?
- In your *ABC* notebook, write a brief report of your findings.
- Based on what you learned, check the visible plumbing in your house or apartment and try to determine what material was used.

LAB 1

HOW IS THE VISCOSITY OF A LIQUID MEASURED?

PREVIEW

Introduction

Mark P. is a plant operator in a polymer production plant. His job is to monitor the quantity of raw materials (monomers) that are added to the polymer reactor as well as the length of time the materials stay in the reactor (reaction time). By changing the quantity of monomer or the time the monomers stay in the reactor, the molecular weight of the polymer can be controlled. If more monomer is added, more monomer is available to react. Therefore, a longer polymer chain with a higher molecular weight can be created. If the reaction time is increased, the monomers have more time to react. This also results in a polymer with a higher molecular weight.

To monitor the molecular weight, Mark sends a sample of the polymer to the lab. The lab technician measures the **viscosity** (vĭ-skŏs′ĭ-tē) of the polymer in a solution. Viscosity is the ability of the polymer solution to resist flow, which is a function of the molecular weight of the polymer. The higher the molecular weight, the longer the polymer chain. The longer the polymer chain, the more polymer interacts with the solvent. The more polymer interacts with the solvent, the higher the viscosity of the solution.

To measure viscosity, the lab technician dissolves a certain amount of the polymer in a solvent. She then measures the time it takes for the polymer solution to pass through a viscometer. A viscometer is a capillary tube with two marks above and below a measuring bulb. The two marks are used as reference points to measure the time the polymer takes to flow through the measuring bulb. Viscosity is related to the flow time of a polymer solution.

Mark knows what the viscosity (and therefore the molecular weight) of the polymer must be for the grade of polymer he is making. The lab results come back showing a viscosity higher than desired. Mark begins the process of determining which factors to change in order to alter the process. Sitting at his computerized control panel, Mark thinks to himself, "Should I change the amount of raw material, or should I change the reaction time?"

Purpose

In this lab, you will test the ability of various liquids to resist flow.

Lab Objectives

When you've finished this lab, you will be able to—

- Rank liquids based on the results of viscosity tests
- Relate the behavior of a liquid to its viscosity in a given situation
- Compare two methods of testing the viscosity of a liquid

Lab Skills

You will use these skills to complete this lab—

- Position a metal object using a magnet
- Time an event using a stopwatch
- Measure distance using a ruler

Materials and Equipment Needed

funnel
viscosity tubes
test tube rack, or ring stand
ball bearing
liquids
marking pen
stopwatch
block or bar magnet
rubber stoppers, solid
bottle cap
tap or distilled water

Pre-Lab Discussion

Viscosity is the degree to which a material resists flow under an applied force. Many oils used in industry are highly viscous, which makes them suitable for applications such as lubricants.

A convenient way to test viscosity is to measure the rate of flow of a fluid through a vertical tube. The time for a given amount of fluid to flow through the tube is an indication of its viscosity. The velocity of a spherical object moving through a fluid at rest is also an indication of the viscosity of the fluid.

Safety Precautions

- If bearings drop on the floor, pick them up immediately so someone does not slip on them and be injured.

LAB PROCEDURE

Method

Put on your lab apron.

Work in groups of four students.

1. Using a marking pen, place two lines 30 cm apart on each of the viscosity tubes as shown in Figure L1-1. Then place a third line 2 cm above the upper line.

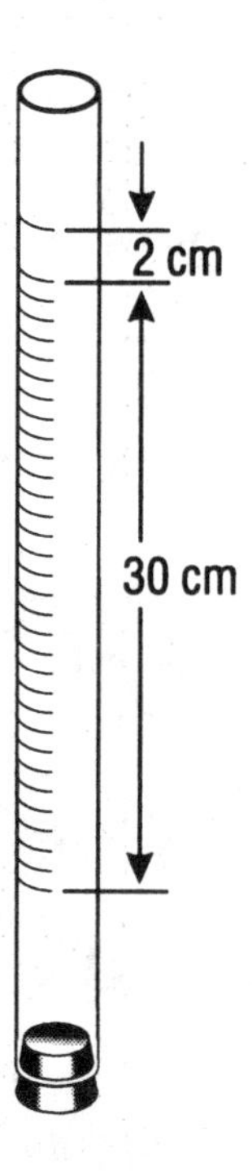

Figure L1-1 Marking the viscosity tube

2. Using a funnel, fill each tube to within 2-3 cm of the top with a different liquid from the lab counter. Fill another tube in the same way with tap or distilled water.

3. Each of the four lab partners will have a specific task during viscosity tests:

    - Partner 1—you will drop the ball bearing into the tube and, using the magnet, maneuver the bearing into position just above the top line on the tube (see Figure L1-2). When your group is ready to start the test, you will allow the ball bearing to drop by pulling the magnet away from the side of the tube.

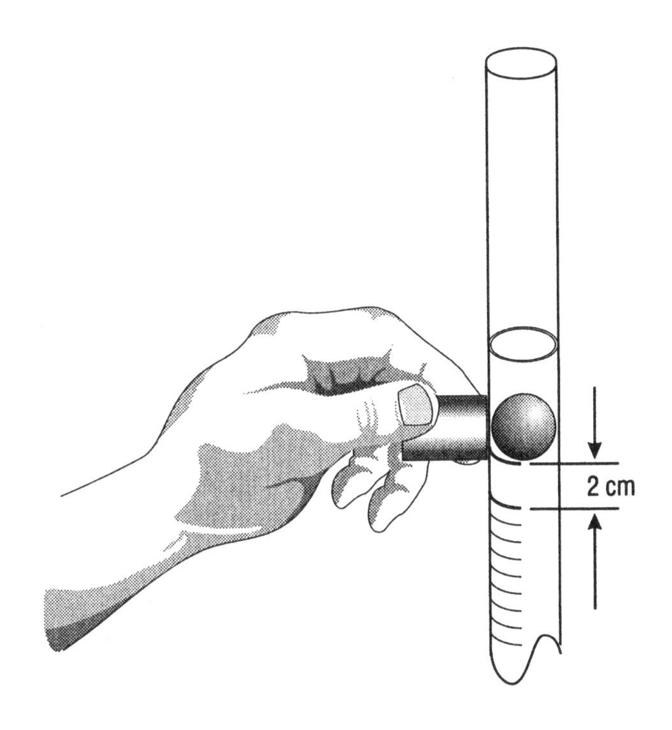

Figure L1-2 Positioning the bearing in the viscosity tube

    - Partner 2—you will reset the stopwatch to 00:00 seconds and move to eye level with the middle mark on the tube. When the falling bearing reaches the top of the second mark, start the stopwatch. Partner 3 will tell you when to stop the watch.
    - Partner 3—you will move to eye level with the bottom mark on the tube. When the falling bearing reaches the bottom line, say, "Stop!" to Partner 2, who will stop the stopwatch.
    - Partner 4—you will read the stopwatch and record the time of fall in the data table.

4. Decide which task each partner will take and carry out the test (Trial One).

5. Repeat the test on the same liquid three times (Trials Two, Three, and Four), with each partner rotating to a new task.

6. Carry out viscosity testing on all the other liquids following the above steps exactly. Clean the ball bearing between tests and keep it in the bottle cap.

Cleanup Instructions

- Return viscous liquids to their stock containers.
- Wash tubes well with soap and water, rinse, and store.
- Clean bearings with a wet paper towel, dry, and store.

Observations and Data Collection

Liquid	Time for the Bearing to Fall 30 cm (sec)				
	Trial One	Trial Two	Trial Three	Trial Four	Average

Calculations

A. For each liquid, compute the average time for the bearing to fall in the trials. Use the formula below.

$$\text{Average time (sec)} = \frac{\text{Total time for the trials (sec)}}{\text{Number of trials}}$$

B. Calculate the relative viscosity of each liquid (except water) using the following formula:

$$\text{Relative viscosity} = \frac{\text{Average time for bearing to fall (sec)}}{\text{Average time for bearing to fall through water (sec)}}$$

WRAP-UP

Conclusions

1. How does the velocity of the falling bearing relate to the viscosity of the liquid in the tube? Explain.
2. Rank the liquids you tested from most viscous to least viscous.
3. Based on your observations and calculations in this lab, decide how a liquid's viscosity would contribute to its ability to
   - serve as a hydraulic fluid
   - rapidly fill a sealed space
   - serve in a shock absorber

Challenge

4. With the viscosity tubes still filled with liquid, place a rubber stopper in each. (Be sure the stopper fits snugly. If it tends to loosen when you push it down, pour out a few milliliters of liquid.) Quickly turn the tube upside down and observe what happens. Use this observation to develop a second method for testing the viscosities of all the liquids. Carry out these tests.

   When you have finished the tests, redo Calculations A and B and discuss the results with your lab group. Consider these questions in your discussion:

   - Why did the ball bearing and the bubble formed when the liquid was turned upside down move in opposite directions?
   - How did the different behaviors of the bearing and the bubble in the liquids influence the outcome of the viscosity tests? Explain.
   - Which of the two viscosity tests (ball bearing or bubble test) do you think is a more accurate indication of the relative viscosities of the liquids? Justify your answers based on your data in each test.

LAB 2

HOW ARE CRYSTALS GROWN?

PREVIEW

Introduction

It was hard for the students visiting Mr. T's Rock and Gem Shop to believe that what they were seeing sparkling from beneath the glass cases were once nothing more than pieces of rock and stone. They were all relieved when Ms. Li asked Frank T., the shop owner and a certified gemologist, what many of them were already thinking.

"Mr. T., aside from their beauty, what makes rubies, emeralds, and opals so much more valuable than other stones such as granite, slate, and feldspar?"

Mr. T. couldn't wait to reply. "First, rubies, emeralds, opals, and of course diamonds, occur much more rarely in Earth's crust than the other rocks you mentioned. But that is only half the story. It's the history of the rocks from which gems come that holds the clue to their value."

Mr. T. continued, "Most gems were formed when the rocks that held them were subjected to very high temperatures and pressures in the Earth's crust. As the rocks cooled, the gemstones crystallized from the molten (hot liquid) rock around them. The conditions under which the rocks cooled helped shape the crystal structure of the gems. That crystal structure really accounts for the gem's effect on us. The unique shape of the crystal of this quartz I'm holding, really just a form of sand, is what determines the colors of light it reflects and how brightly it sparkles."

Mr. T. went on to tell the class that the heart of any gemologist's work is knowing what stones to select and how to cut and polish them to emphasize their natural crystal shape. Mr. T. put his jeweler's lens in place and began polishing a rock from a recent collecting trip. As the class watched him work, Ms. Li seemed very satisfied. It was as if she had known all along that a trip to Mr. T's gem shop would be a great way of introducing them to the chemistry of crystals.

Purpose

In this lab, you will grow crystals of two different salts from saturated solutions.

Lab Objectives

When you've finished this lab, you will be able to—

- Grow crystals of a salt from a saturated solution
- Evaluate conditions that determine the size of the crystals

Lab Skills

You will use these skills to complete this lab—

- Use a hand lens to magnify an object
- Dissolve a solute in a solvent
- Weigh a substance using a balance

Materials and Equipment Needed

alum
sodium chloride
triple-beam balance
deionized water
6 petri dishes
funnel, small plastic or glass
string
small washer
safety goggles
100-ml graduated cylinder
2 600-ml beakers
hot plate
wax pencil
2 glass stirring rods
magnifying glass
hot pad
lab apron
tall glass cylinder (such as hydrometer cylinder)

Pre-Lab Discussion

A substance in its liquid phase has no definite shape—it simply takes the shape of whatever container it may be in. One way to explain this behavior of a liquid is to say that the molecules of the liquid have enough kinetic energy to overcome the forces of attractions among them. This energy means that the molecules are not in a fixed position with respect to one another. Instead, they are free to move around one another so that the liquid itself is fluid—it flows.

When a liquid is cooled, the molecules lose kinetic energy and can no longer overcome the forces of attraction. The molecules become arranged in a more orderly fashion. Some substances, when cooled under the right conditions, will form crystals. Crystals of compounds known as salts (for example, sodium chloride or aluminum sulfate) can be formed from a saturated solution of the salt in water. (A saturated solution is one in which no more solid substance, or solute, can be dissolved.)

In making crystals in this lab, you will make use of the effect of temperature on how well a solid substance dissolves in a liquid—in other words, how temperature affects solubility. As Figure L2-1 shows, increasing the temperature of the solution will increase the solubility of most solids. This principle means that, when a solution is heated, you can get more solute to dissolve in it. But it also means that, when the solution is cooled, it must "give up" some of its solute.

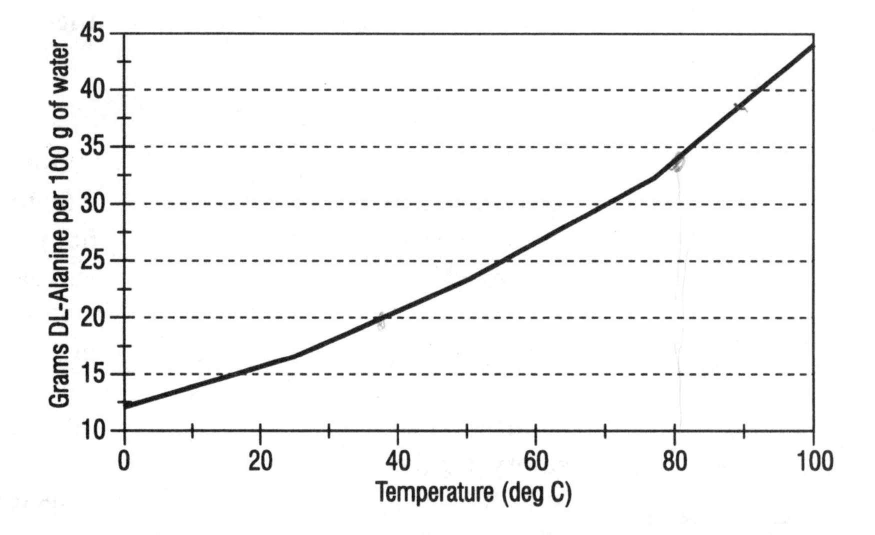

Figure L2-1 Solubility depends on temperature.

If the cooling of a saturated solution is strictly controlled, crystals of a certain shape and size will form. In making crystals in this lab, you should be aware that the cooling of a solution is a result of two factors under your control:

- The net loss of heat by the solution to its surroundings, which begins as soon as you remove the heat source.
- Evaporation of the solvent (in this lab, water) helps speed the rate of heat loss by the solution and increases the concentration of the solute.

In this lab you will have the opportunity to experiment with different ways of controlling the formation of crystals from a salt solution. If you get a crystal formation that you are especially proud of, your teacher will show you how to preserve it so that you can show your friends outside of class.

Safety Precautions

- Use a hot pad, and be careful when removing the beaker from the hot plate. Spilling the hot liquid on clothing or skin can cause painful burns.
- If you are burned, run cool water over the burn and report it to your teacher immediately.

LAB PROCEDURE

Method

Put on your lab apron and goggles.

Part A. Preparing the Saturated Salt Solutions

1. With the 100-ml graduated cylinder, measure 400 ml of deionized water into one 600-ml beaker and 200 ml of deionized water into the other 600-ml beaker. Use the wax pencil and label the beaker with 400 ml of water "alum" and the beaker with 200 ml of water "NaCl."
2. Place the beakers on a hot plate and heat the water to boiling. While the water is heating, weigh 100 g of alum and add it to the beaker labeled "alum" and weight 80 g of NaCl and add it to the beaker labeled "NaCl." Stir the solutions with separate glass stirring rods. Use only one stirring rod in a solution and do not use the same stirring rod in more than one solution. Stir the solutions until all the solute is dissolved or the solution boils for 2 minutes.

Part B. Encouraging Crystal Formation Using a Substrate

3. Use a glass stirring rod, string, small washer, and tall cylinder arranged as shown in Figure L2-2.

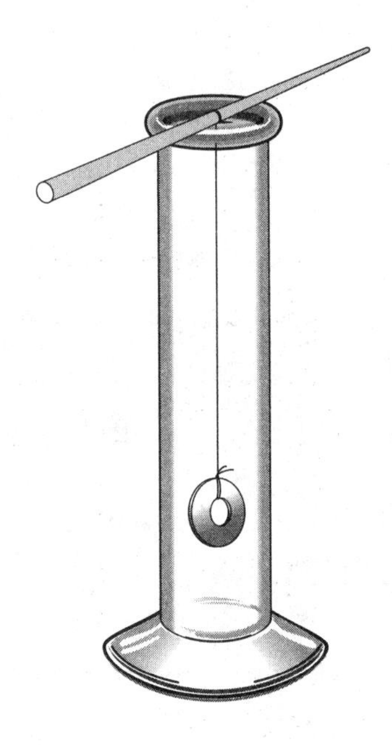

Figure L2-2
Arrangement of cylinder

4. Remove the washer and string from the cylinder. Using a funnel, carefully pour the hot alum solution into the cylinder until it is about 3-4 cm from the top of the cylinder. Replace the string and washer. Place out of the way on the lab table.

Part C. Controlling Formation of Crystals

5. In your *ABC* notebook, list at least three ways you can vary the conditions for cooling the saturated solutions to form crystals. You must form your crystals inside the petri dishes provided. Include one dish with alum and one dish with NaCl for each cooling condition you will test. Use the wax pencil to label the bottoms of the petri dishes with appropriate labels. Place the four petri dishes on a level area where they will not be disturbed. Pour the hot solutions to a level near the top of the dishes. Place the covers on any dishes that are supposed to be covered.

6. Incubate dishes at the desired temperatures. After one week, observe the contents of the dishes and the cylinder. Use a magnifying glass to observe any crystals that have formed. Record your observations and sketch the crystals in the Data Table.

Data Table for Parts B and C

Dish	Condition (describe)	Observation	Sketch of crystals
Cylinder with alum			
Alum			
NaCl			
Alum			
NaCl			
Alum			
NaCl			

Cleanup Instructions

- Empty unwanted crystals into the trash can.
- Pour any remaining salt solution into the sink.
- Wash the glassware and return it to its proper location.

Calculation

No calculations are needed for this lab.

WRAP-UP

Conclusions

1. Use Table 1-1 to identify the crystal system to which your crystals belong.
2. Compare the crystals of alum that grew in different cooling conditions. Which condition grew the largest crystals? Make this same comparison among the NaCl crystals. Explain why crystals grew as they did under each condition.

Challenge

3. Give an explanation for the difference in the size of the alum crystals grown on the string and the alum crystals grown in the petri dishes.

SUBUNIT 2

METALS AND THEIR ALLOYS

THINK ABOUT IT

- What parts of a car are made of metal and why?
- Why doesn't a metal sink get rusty the way a shovel or hoe does if it's left out in the dew or rain?
- Why are bridges made of steel?
- Why are electrical wires made from copper?
- Why doesn't a gold ring turn your finger green the way a brass ring does?

SUBUNIT OBJECTIVES

After you complete this subunit, you will be able to—

1. Relate the physical and chemical properties of metals and alloys to their uses.
2. Explain the properties of metals and alloys in terms of atomic, crystalline, and grain structure.
3. Predict the properties of steel alloys based on the processes used to make them.
4. Characterize metals based on their performance in tests for certain properties.
5. Compare two different methods for testing the hardness of a metal.
6. Rank a series of metals in order of their hardness.
7. Identify chemical change on a metal sample.
8. Predict the outcome of connecting two metal samples.

PROCESS SKILLS

You will use these skills in lab—

- Use a file to scratch metals.
- Use a ruler to measure diameter.
- Use a magnifying lens to aid in taking measurements.
- Measure voltage with a digital voltmeter.
- Make electrical connections between two metal samples.
- Observe metal samples for evidence of chemical change.

What Are Metals and What Makes Them Useful?

Metals are among the most useful materials available to humans. They have played such a significant role in the development of civilization that periods of development have been named after the metal most used in that era—the bronze age or the iron age, for example.

What about metals makes them so important? How are they used today?

Physical Properties of Metals

CAREER PROFILE: WELDER/OWNER OF A METALS FABRICATION BUSINESS

Skot I. is the owner and manager of Skotz, Inc., a steel fabrication company. The company specializes in building metal structures for wastewater-treatment facilities and oil field equipment. Both types of structures are generally very large. "When you talk tons, that's when I get interested," says Skot. He goes on to describe a recent custom product: a flare stack at an oil field facility that is 200 feet tall, 8 feet in diameter at the base, 88,000 pounds in weight, and able to resist winds of up to 120 miles per hour. "That gives you an idea of the scale we're into," he says. (By contrast, most cars weigh between 2,000 and 3,000 pounds.) "The main part of the stack was made of carbon steel, but the top of the stack—the area where escaping gases burn in the air—was made of a stainless steel called incolloy."

Skot described welding oil field equipment as demanding. "A problem weld could cause an explosion. For everything we do—the water treatment work included—there are fabrication codes and third-party inspectors who X-ray the welds to verify their quality."

When he's asked how long it takes to become a good welder, Skot first responds, "Another twenty or thirty years and I should be there," but he goes onto distinguish between different skill levels. "Being a welder is fairly easy, if you are doing the same type of weld every day and someone is showing you where to weld. You can learn that in a few weeks. But that type of welding is being done more and more by robots. Lots of welding in the automobile industry is now done by robots."

"Being a welder who can read and follow blueprints, measure accurately, and maintain a high skill level while working upside down or at a difficult angle—that takes a long time, a lot of experience. That kind of welder will also be in demand for a long time to come."

When he is asked what is the most satisfying aspect of welding, Skot answers readily. "Building something that will be there for a long, long time—that's satisfying. Who wouldn't be proud to have had a part in making a beautiful steel bridge, for example?" Although his company doesn't build bridges, it does make equipment that will, in Skot's words, "be there long after we're gone."

The welder and business owner described in the preceding career profile uses metals primarily for their structural properties—to make large-scale containers, piping, and other equipment. He has learned about the various physical properties of different metals such as strength and heat conductivity and heat resistance, to name a few.

Activity 2-1

- Bring one or more metal objects (screws, nails, clocks, pens, tools, and so on) to class with you.
- Working in groups of six to eight students, try to identify the kind of metal used in each object. Your teacher will give you a handout of clues to use in metals identification.
- Let each member of the group look up information about one of the metals that you identified (based on educated guessing) and give a brief report to the group.
- As a group, try to decide why the objects were made of the particular metals used. What properties do the different metals have? What properties do the different metals have in common?
- Record your findings in your *ABC* notebook for use in the Unit Wrap-Up Activity.

As you discovered in Activity 2-1, metals have a number of properties in common. They are good conductors of heat and electricity. They can be drawn into wires (they are ductile). They can be beaten into sheets (they are malleable). They are opaque but can be polished to efficiently reflect light (they have a metallic sheen). These are the properties or the characteristics that classify metals as metals. What causes metals to have these properties?

Metallic Bonding—A Key to Understanding Metal Properties

Bonding in metal—metallic bonding—is modeled simply by the example in Figure 2-1. In this model, the metal atoms in a crystal are represented as metal ions (plus signs in the figure). Positive and negative charges are attracted to each other. Two positive charges repel each other; two negative charges repel each other. The attraction and repulsion of electrical charges for each other cause every kind of bonding that occurs between atoms or molecules. In the crystal in Figure 2-1, there is an electron for each positive charge in each ion. This gives the material an overall neutral charge—that is, neither positive nor negative.

The cloud of electrons shown in Figure 2-1 does not lock the metal ions into any fixed position. Therefore, the metal ions can move around to some extent. This limited mobility of the metal ions explains many metallic properties, while the cloud of "free" electrons explains many others.

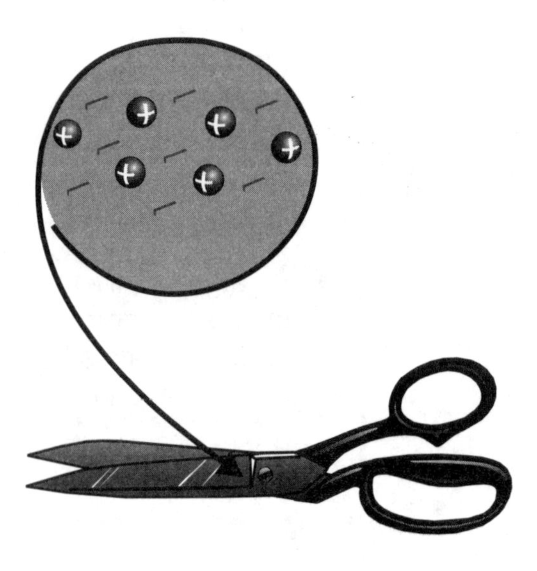

Figure 2-1 Metallic bonding

The cloud of "free" electrons exists because the metal atoms have large regions of space where the electrons can be at a given moment. Recall that the regions of three-dimensional space are called orbitals. Each orbital can hold a maximum of two electrons. When the metal atoms form crystals, these large outer orbitals overlap and form large regions of space from combinations of all of the overlapping orbitals. In other words, the electrons in the outer orbitals are shared among many atoms.

The orbitals are occupied according to how much energy an electron possesses when in that orbital. The electrons are most stable when they have the least energy. This means that electrons, like people, are least likely to move when they have little energy. Therefore, the orbitals in which electrons have the least energy are occupied first.

How Does Electrical Conductivity Result from Metallic Bonding?

The relatively close energy values of the combined orbitals in metals give a narrow range of energies for the electrons in these orbitals. The close energy values mean that an electron can easily go from one orbital to another unoccupied orbital. This narrow range of energies, or band, has a special name. It is called a conduction band because it permits current flow by the movement of electrons from one orbital to another.

Although each of the combined orbitals can hold two electrons, these combined orbitals are so close to the same energy in metals that there may be only one electron in each orbital. This is especially true when a voltage difference is applied across the metal and an electrical current is flowing as the electrons move from one orbital to another.

The combination of orbitals occurs in many crystals. However, it does not always give rise to a narrow energy range as in metals. There are three classifications of material, based on ability to conduct an electrical current:

- Conductors (like the metals)
- **Semiconductors** (like metaloids—the elements along the line that separate metals and **nonmetals**)
- **Insulators** (like nonmetals)

Semiconductors and insulators do not have the narrow energy range that metals do. These materials have an energy gap between the occupied orbitals and the empty orbitals that can permit the flow of an electrical current. The gap is large in insulators and much smaller in semiconductors. These energy ranges or bands are shown in Figure 2-2.

The existence of the overlap region or the separation of the bands (size of the band energy gap) depends on the orbitals that form the combined orbitals. Before a semiconductor or insulator can conduct a current, the electrons must have enough energy to jump the band energy gap. This energy is supplied by a voltage or potential energy difference across the material. Low voltages are usually adequate for semiconductors to conduct a current, but very high voltages are needed for insulators. When an insulator conducts under high voltage, it is said to break down.

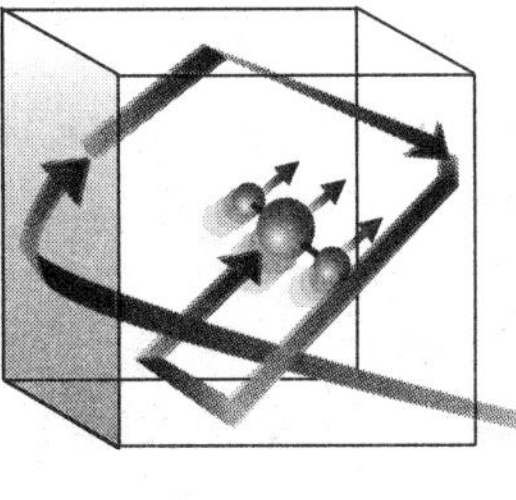

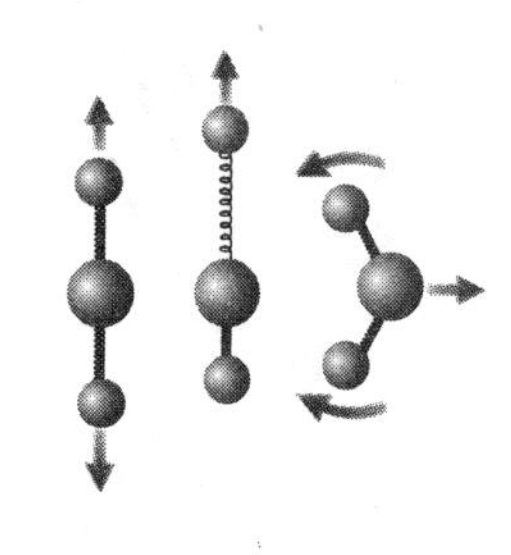

a. Translational b. Rotational c. Vibrational

Figure 2-2 Energy ranges or bands

How Does Heat Conductivity Result from Metallic Bonding?

As already discussed, metals are good conductors of electricity. We have said that this is because of the ease with which electrons can move around in the material. Metals are also good conductors of heat. In this case, too, metals use the mobility of their "free" electrons.

Heat energy is stored in a material as molecular or atomic kinetic energy—that is, in the motion of its molecules. The types of molecular motion are

- **Translation**—movement from one position to another in a straight line

- **Rotation**—movement of the molecule around its center of mass
- **Vibration**—movement of the atoms in a molecule or crystal relative to each other

As a material heats, the kinetic energy of the molecules increases. This means that, for a gas or a liquid, the molecules move back and forth faster (translate), rotate faster, and vibrate faster. For solids, the only movement the molecules can have is vibration. Therefore, when a solid heats, the molecules of the solid vibrate faster.

Heat energy moves from a higher temperature to a lower temperature. This means that, if a hot gas and a cool gas are connected by a tube as shown in Figure 2-3 and the molecules are constantly colliding, hot gas molecules—moving rapidly—and cool gas molecules—moving slowly—will collide. The collision transfers energy from the fast, hot molecule to the slow, cool molecule. This transfer of energy slows—and cools—the hot molecule and at the same time increases the speed of—and warms—the cool molecule.

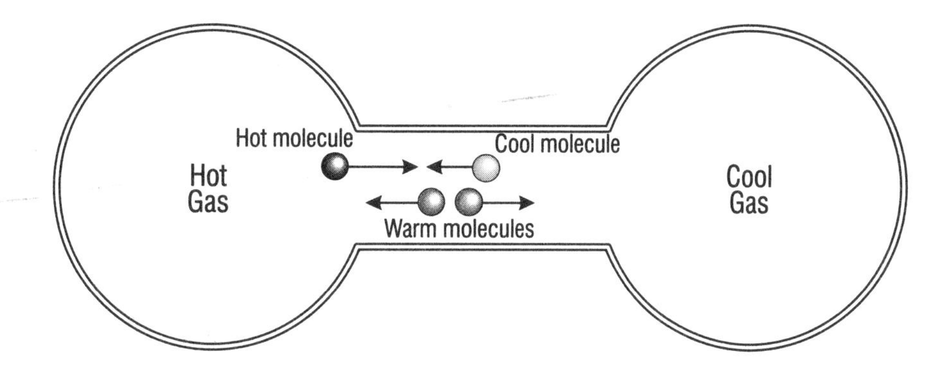

Figure 2-3 Heat conduction by a gas

Why are metals such good heat conductors? Most metals at room temperature are solid materials. This means that crystal vibrations of the atoms are the only method of storing heat energy. As the atoms vibrate faster as a metal heats, the atoms collide with the drifting "free" electrons. This collision transfers energy to the drifting electron, allowing it to translate faster. Therefore, the "free" electrons in a metal transfer heat in much the same way the gas molecules do in Figure 2-3.

Activity 2-2

- While metals are good conductors of heat, some metals are better conductors than others. In this activity, you will compare the heat conduction of four metals to each other and to plastic, a thermal insulator.
  - Work in groups of four.
  - Get a heat conduction kit from your teacher.
- Follow the instructions included in the kit. Caution—be careful with the open flame.

How Does the Formability of Metals (Malleability and Ductility) Result from Metallic Bonding?

Most metals can be formed either by hammering the material into sheets (malleability) or by drawing the material into wires (ductility). These properties of metals are used to form metals into useful objects.

A process called "pressing" is often used to give metals a specific shape (see Figure 2-4). Cake pans and automobile hoods are examples of products made by pressing. Items such as these are made by placing a sheet of metal in a form called a **die**. The die has the shape of the final product, just as a cookie cutter has the shape of the cookies it makes. When the die and the metal are pressed together, the metal deforms and takes on the shape of the die. Only metals that have the property of malleability can be processed through pressing.

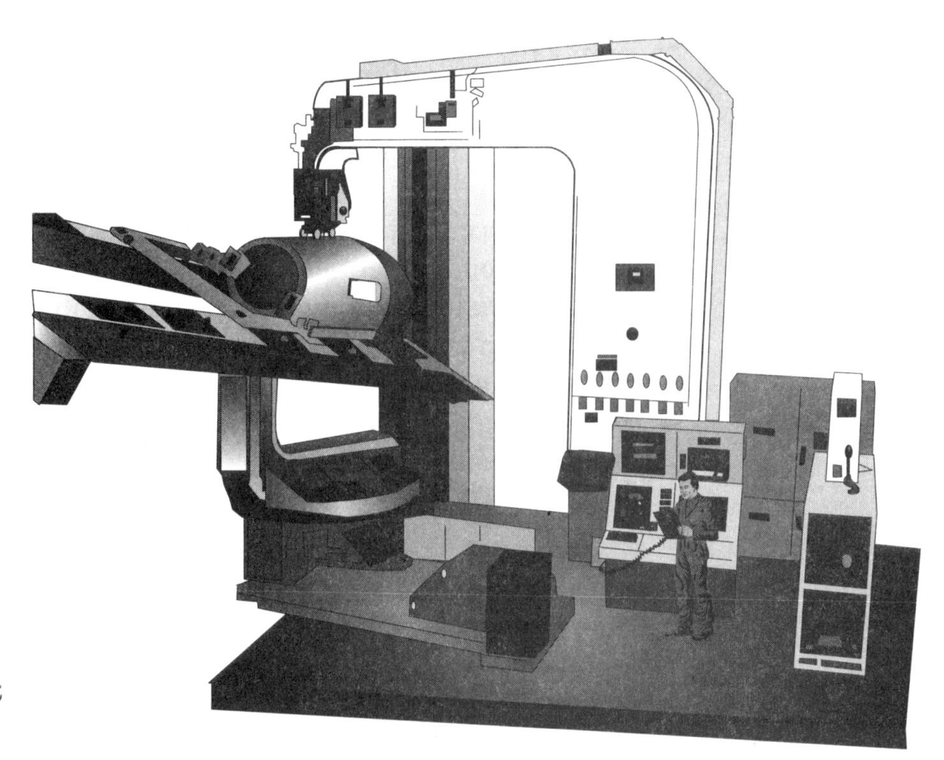

Figure 2-4
Forming sheet metal parts

Ductility is another characteristic of metal. See Figure 2-5. The term ductility refers to the ability of a metal to be drawn into a wire or rod. (Like malleability, ductility requires that a material be able to be plastically deformed without breaking.) Coat hangers and paper clips are examples of products made of metals having the property of ductility.

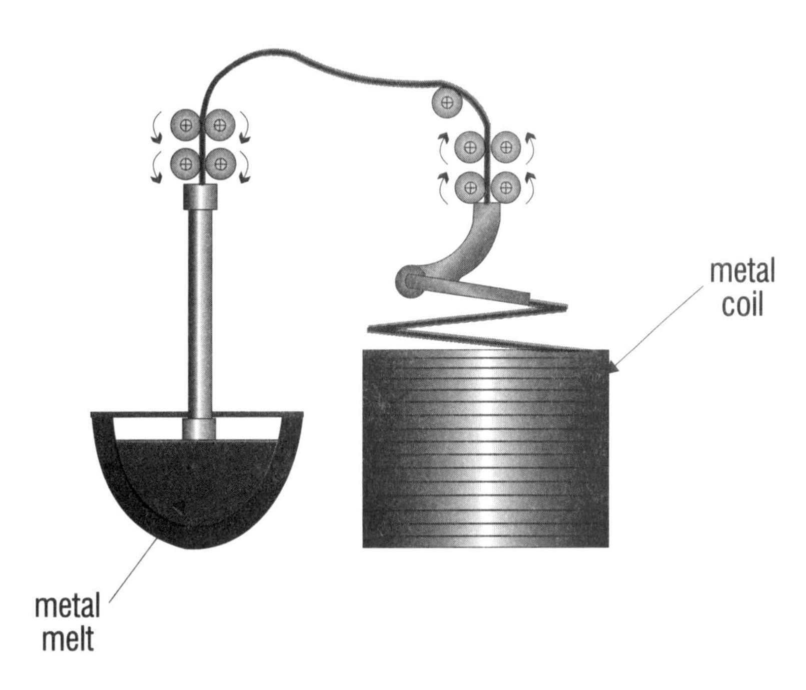

Figure 2-5
Extruding wire

The nature of metallic bonding helps to explain the properties of malleability and ductility just as it helps to explain electrical conductivity and heat conductivity. The metal atoms form the combined orbitals and the outer electrons occupy the combined orbitals, thus creating the cloud of mobile "free" electrons. The remainder of the metal atom is considered a metal ion. The mobile electrons do not rigidly hold these metal ions in the **crystal lattice**. Instead, the material is similar to chocolate chips (metal ions) in cookie dough (mobile electrons) and it can be formed easily into the desired shapes.

Chemical Properties

Late Night Briefing

The campaign appearance was almost over. The senator had time for one more question from the crowd, and a heavyset man in a plaid workshirt had just stood up to ask it. "Senator, what are you planning to do about the ***infrastructure*** (ĭn′frə-strŭk′chər) *?" the man challenged. "I've read that some 200,000 highway bridges in the United States are deficient now. I'm not sure exactly how deficient, but I don't think we should wait around to find out."*

"I'm not sure where that number comes from," said the Senator, "but we are aware of the problem. The president has included some transportation solutions in his third-year economic stimulus package and we'll be looking at that as soon as it's off the drawing boards."

With that, the senator was stepping off the podium, shaking hands as she moved toward the door. Much later, in her hotel room, she could be heard putting in a call to Maria Martinez, her technical advisor in Washington.

"Miss Martinez, I was on the spot today, and I want you to fill me in—are there really 200,000 deficient highway bridges in the United States?"

"That sounds about right." Maria Martinez was calm on the other end of the line.

"Well, what the heck's wrong with them, and what are we doing about it? Aren't these bridges made of steel and concrete? What could cause problems with that?"

"It's a bit complicated, Senator. One of the major problems is corrosion of the steel reinforcing bar—rebar—that is used

in these steel and concrete bridges," began Maria. "Salt spray from seawater and deicing compounds are the big culprits."

"Wait a minute," came back the senator. "Doesn't concrete protect the rebar?"

"Yes, if it's thick enough and if the salt solution is not too concentrated. But there are lots of bridges with exposed rebar or thin concrete. And with a salt solution of more than 250 parts per million . . ."

"Miss Martinez, I've asked you before, please don't talk in numbers; it's not my first language."

"Sorry, Senator. What I'm saying is that seawater and the runoff from a deiced bridge are such strong salt solutions that the concrete doesn't provide protection against corrosion. The bridge acts like a giant battery. There sits the concrete soaked with an **electrolyte** (ĭ-lĕk′trə-līt′) *solution, in this case, salt. The rebar acts like a cathode in the battery. It's the terminal that carries a negative charge and it's where oxidation occurs . . ."*

"And what does that mean?" the senator broke in impatiently.

"Right, senator. That means that corrosion begins to occur in the rebar just the way a battery terminal gets corroded because of oxidation reactions taking place there. So the bridge eventually gets weaker."

"So, what are we supposed to do about this? Because I'll tell you what we can't do. We can't rebuild 200,000 bridges."

"Fortunately, there is a new solution that seems to be working well," explained Miss Martinez. "It involves spraying the bridges with a zinc coating. Under salt conditions, the zinc becomes the **cathode** *where oxidation—and therefore corrosion—takes place. The rebar becomes the* **anode**. *As an anode, it no longer undergoes corrosion."*

"And the cost of zinc spray?" asked the senator.

"More than paint coatings to begin with, but more than 25% less than painting them in the long run," Maria replied.

"Okay, Miss Martinez, that's about all the science I can learn in one phone call. Can you fax me something about this tomorrow?"

"Sure, Senator, it's on its way. Goodnight, Senator."

"Goodnight, Miss Martinez."

Maria Martinez has a challenging job—explaining a very technical problem to a decision maker without a technical background. She must explain in everyday language to the senator what happens to the steel rebar in the bridges. In other words, she needs to convey technical ideas in non-technical language. To do that, Maria must understand the chemical properties of the metals.

The chemical properties of metals are determined by the arrangement of the electrons in the atom. Only a few of these electrons are actually involved in chemical reactions. In most cases, the only electrons that take part in chemical reactions are the outer electrons—called **valence electrons**. An atom can do three things during a chemical reaction:

- Gain electrons completely—be reduced to negative ions
- Lose electrons completely—be oxidized to positive ions
- Share electrons—form a covalent bond with another atom

When metals undergo chemical reactions, they are usually oxidized. They lose electrons and become positive ions. However, under certain conditions, metal ions can be reduced (gaining electrons). When the steel rebar in the bridges that have been treated with zinc comes in contact with the salt solution, the iron ions in the rebar are reduced to become iron atoms again. At the same time, the zinc atoms are oxidized to become zinc ions. This kind of reaction is called an oxidation-reduction or redox reaction. The reactions involved in this process are

- without zinc protection—

  oxidation of iron to ferrous (iron II) oxide

$$\underset{\text{(iron)}}{2\,Fe} + \underset{\text{(oxygen)}}{O_2} \longrightarrow \underset{\text{(ferrous [}Fe^{+2}\text{] oxide)}}{2\,FeO}$$

Equation 2-1

oxidation of ferrous (iron II) oxide to ferric (iron III) oxide

$$\underset{\text{(ferrous oxide)}}{4\,FeO} + \underset{\text{(oxygen)}}{O_2} \longrightarrow \underset{\text{(ferric [}Fe^{+3}\text{] oxide)}}{2\,Fe_2O_3}$$

Equation 2-2

- with zinc protection—

  oxidation of iron to ferrous (iron II) oxide

  $$2\,Fe + O_2 \longrightarrow 2\,FeO$$

  (iron) (oxygen) (ferrous [Fe^{+2}] oxide)

  Equation 2-3

  oxidation of zinc

  $$Zn + O_2 \xrightarrow{H_2O} Zn^{+2} + 2\,OH^-$$

  (zinc) (oxygen) (in water) (zinc ion) (hydroxide ion)

  Equation 2-4

  simultaneous reduction of ferrous (iron II) oxide to iron

  $$2\,FeO \xrightarrow{H_2O} 2\,Fe + O_2$$

  (ferrous [Fe^{+2}] oxide) (in water) (iron) (oxygen)

  Equation 2-5

Reactivity of Metals

Many metals, because they lose electrons so easily, are so reactive that they are not useful for bridges or other types of structures. Metals like lithium, sodium, and potassium are so reactive that they react with water. Lithium bolts in a bridge, for instance, would oxidize the first time they came in contact with water and would even react with the water vapor in the air.

$$2\,Li + 2\,H_2O \longrightarrow 2\,LiOH + H_2$$

(lithium) (water) (lithium hydroxide) (hydrogen)

Equation 2-6

Such very reactive metals are stored under kerosene or some other nonwater solvent to keep water away from them. Because of their reactivity, these metals are not useful in structures such as bridges or piping, but they have other uses based on their reactivity. For example, lithium is used in lightweight batteries with long shelf lives.

CAREER PROFILE: BIOMATERIALS SAFETY MANAGER

Kyle W. works for a company that make surgical instruments as well as materials that are implanted into the body such as **synthetic** heart valves. Kyle doesn't talk about specific products or processes when he talks about his job—partly because of his company's guidelines about giving out information. Instead, he describes in a general way the area he manages.

"My primary responsibility is sterility testing," Kyle says. "My function is to ensure that all the products sold as sterile are indeed sterile." Kyle goes onto explain that companies that make medical devices—from bandaids to breast implants—must follow guidelines set by the Food and Drug Administration (FDA).

"All medical devices are classified according to factors such as where they are used, how they are used, and how long they are used," explains Kyle. "The FDA sets guidelines based on the classification. Our company has to make sure that our products meet FDA guidelines. Other people in the company consider physical properties of materials such as their strength or flexibility. My job is to make sure that we meet the FDA guidelines for our product in the area of **biocompatibility**."

"For a material that comes in contact with internal body tissues or that is implanted inside the body to be considered biocompatible, it must be safe for long-term use. That means it cannot have any of these effects: systemic (affecting body systems), cytotoxic (affecting the body's cells), **carcinogenic** (kär-sĭn′ə-jənĭk) (cancer-causing) or genotoxic (affecting the offspring of the person)."

To ensure sterility, Kyle and his coworkers carefully monitor the processes used to make the medical products. They also take samples of the products after they are made and test the samples to be sure they are sterile.

Besides monitoring and testing for **sterilization** (stĕr′ə-lĭ-zā′shən), Kyle tests any materials that may be considered for use in a product or that might come in contact with that product. "A material that is used to clean a mold that is used to make one of our products has to be tested to make sure that it won't have adverse effects on the product itself," says Kyle. "If a plastic bag is going to contain the product for even a short time, we test that, too."

As Kyle talks about his job, it becomes clear that he feels very close to the patient side of what he does, "Some of our products are life-sustaining for the people who use them," he says. "I like to be involved with the production of devices that help people." Kyle has a bachelor's degree in microbiology and a master's degree in cellular biology.

Using the Periodic Table to Find Out about Metal Reactivity

How do people like Maria Martinez know what metals are highly reactive and what metals are less reactive? The periodic table offers a lot of help with this task. Lithium (Li), sodium (Na), and potassium (K) are all Group 1 elements (in the first column of the periodic table), each of which has a single electron in the ns orbital. This location in the periodic table means these metals have only one valence electron; that is, each has only one electron in its outer orbital.

Because the single electron is easily lost, these elements are reactive metals. Elements in Group 1 are often called alkali metals. They react violently with water, giving off hydrogen gas and forming a metal hydroxide solution, which is an alkaline solution. The alkali metals are soft enough to be cut with a knife. The freshly cut surface of an alkali metal is shiny but becomes dull quickly as it is oxidized by the oxygen in the air. Alkali metals have low densities, low melting points, and good electrical conductivity.

The second column of the periodic table is Group 2. The elements in this group have two electrons in their ns orbitals. The Group 2 metals are also very reactive, but not as reactive as the Group 1 metals. The elements in this group are called alkaline earth metals because they also form alkaline solutions. Earth's portion of the name comes from their being found in ores in earth.

Newly exposed surfaces of the alkaline earth metals react quickly with the oxygen in the air to form a thin oxide coating. The oxide coating protects these metals from additional oxidation. Protection from further oxidation makes the metals useful in structures where light weight and high strength are required.

Activity 2-3

- Working in pairs, get a 3-inch piece of magnesium ribbon, a piece of sandpaper, and a 250-milliliter beaker half filled with water from your teacher. Sand about one inch of one end of the magnesium ribbon. Immerse the ribbon in the water in the beaker. Observe the two ends of the ribbon for ten minutes.
- Describe in your *ABC* notebook the differences in what you see on the two ends of the ribbon.

Many of the metals used in alloys (materials in which metals are combined with one another or other elements) are called **transition metals**. This means they are located in the periodic table between Group 2 and Group 13 and are in the fourth, fifth, or sixth period (row). The groups in the transition metals are Groups 3 through 12.

Transition metals (and lanthanides and actinide) are the only elements in which electrons below the valence level (outer level) participate in chemical reactions. Beginning in the fourth period, potassium (K) and calcium (Ca) fill the first orbital in the fourth level (called the 4s-orbital). Then, with scandium (Sc), the electrons drop back to the third level and fill orbitals called d-orbitals throughout the transition metals. The d-orbitals in the process of being filled are always one level lower than the valence level. However, these d-electrons do enter into chemical reactions.

In Subunit 1, you learned that the known elements are divided into two main groups—metals and nonmetals. Some elements in the metals group are well known for their useful properties. Examples are aluminum (Al), silver (Ag), iron (Fe), and copper (Cu). These metals are very useful partly because they are among the most durable and least reactive of metal elements. However, although the metals are not so reactive as to be considered unstable, they are capable of combining with other metals to make the materials known as alloys.

What Are Alloys and What Makes Them Useful?

An alloy is produced when a metal is combined with another metal or a nonmetal to produce a new material. For example, steel is an alloy of iron (Fe) and carbon (C). Brass is an alloy of copper (Cu) and zinc (Zn). Bronze is an alloy of tin (Sn) and copper (Cu).

Alloys Are Solid Solutions

Most alloys are solutions formed of one or more elements dissolved in metal. A solution is a mixture in which one substance is thoroughly dissolved in another. You can see a solution formed when you put sugar in iced tea. However, if you put in more sugar than can be dissolved, some sugar comes out of the solution and settles to the bottom of the glass.

The term "solid solution" describes a solution in which both of the materials normally occur as solids. Cast iron, an alloy made of iron (Fe) and carbon (C), is a good example. In a mixture of carbon and iron, carbon does not dissolve at room temperature. At around 843°C (1550°F), both carbon and iron melt in such a mixture. In the molten state, iron readily dissolves carbon. **Cast iron**—so-called because it is formed in a mold in a process called casting—has so much dissolved carbon added to it that some of the carbon settles out in little spheres, like sugar settles out of iced tea. (This carbon helps lubricate tool bits when you drill or machine cast iron.)

Iron (Fe) is such an important ingredient in alloys that they are classified according to their iron content. Alloys are classified as **ferrous** (fĕr′əs) if the major element is iron and **nonferrous** (nŏn fĕr′əs) if the major element is another metal such as aluminum (Al) or copper (Cu).

Activity 2-4

- Using a technical dictionary or encyclopedia, investigate the differences in the following types of iron: pig iron, cast iron, and wrought iron.
- In your *ABC* notebook, write a brief definition of each and an example of how each one is used.

Many Kinds of Steel

One of the most useful alloys is steel. We will discuss it here in some detail because it has so many uses and so many forms. Like cast iron, steel is made primarily of iron and carbon. However, steel has a lower carbon content than cast iron. Most steel is iron with less than one percent of dissolved materials. However, the one percent ingredient in an alloy can have a very powerful effect on the way that steel behaves. To put it another way, the one percent ingredient changes the properties of the steel.

How Does the Carbon Content of Steel Affect Its Properties?

In industry, different steels are divided into families or groups based on their carbon content. The amount of carbon has a direct effect on the properties of the steel. The major groups and some of their uses are shown in Table 2-1.

Table 2-1. Steel Groups Based on Carbon Content

Group	Carbon Content	Examples of Use
Low carbon	less than 0.15% carbon	Wire rivets, nails, welding materials
Medium steel	0.15 to 0.30% carbon	Structural shapes, angles, channels, I-beams
High carbon steel	more than 0.50% carbon	Tools, drills, railroad rails

As the carbon content of steel is increased, its tensile strength is increased. Tensile strength is the ability to resist breaking from pulling forces. However, the ability of the alloy to give or bend without breaking (ductility or malleability) decreases as carbon is added to steel. The formulation of steel always involves a trade-off between these properties—tensile strength on the one hand and ductility on the other.

You can see from Table 2-1 how this change in properties leads to different applications or uses for each type of steel. Why would structural shapes such as I-beams and channels need to have more ductility than drill bits or railroad rails? Which application calls for steel with the greatest amount of tensile strength?

How Does the Addition of Other Elements Affect the Properties of Steel?

New Kid on the Job

Will was happy in his job as an apprentice machinist. It was his first job since he graduated from high school seven months ago. He had liked working with machine tools in high school, so this was a great opportunity to do something he enjoyed and get paid for it. He also found that he knew quite a few of the other employees. The inspector who was checking his first lot of production parts was M.J., a buddy of Will's older brother during their high school days.

Most of Will's first lot of parts passed M.J.'s inspection, but two pieces were slightly undersized. Eager to correct his errors, Will suggested that he'd make two new parts out of a piece of bar stock he'd found behind his lathe. M.J. reaction surprised him. "Hold it, hold it, hold it!" said M.J. "You are too eager, man. You don't even know what that piece of stock is made of? You have to be sure—positive—what alloy you're using. These parts are designed for a very specific use. You can't go substituting one metal alloy for another."

Will felt a little silly, but as M.J. went on, he got more interested in what he was saying and less worried about not knowing it already. After all, M.J. was just trying to help him do the job right, as he explained how different alloys perform. "Remember that off-road bicycle your brother handed down to you? Why do you think it was still in one piece after your brother raced on it for two years? Because it had a 4130 chromoloy steel tube frame, that's why. This bar you found behind your lathe is steel, too, but it's mild steel. There's a world of difference."

"What kind of difference?" asked Will.

"The 4130 steel, if it's heat-treated, can withstand a steady pull of up to 120,000 pounds per square inch (827,000,000 N/m^2). That's its tensile strength. The mild steel bar is good for only 55,000 pounds per square inch (379,000,000 N/m^2)."

M.J. sure seemed to know a lot about the materials they were using. Suddenly it hit Will that he wasn't in school anymore. Now what you knew really mattered, not because you were going to have a test on it, but because you couldn't do your job right if you didn't know it.

Carbon (C) is not the only element that is added to iron to make steel. Alloys of steel often contain small amounts of one or more additional elements such as manganese (Mn), aluminum (Al), phosphorus (P), molybdenum (Me), tungsten (W), silicon (Si), and even gases such as nitrogen (N). Each of these alloys has an effect on the properties of steel. Elements maybe added to gain strength, corrosion resistance, malleability or other properties.

Here we will look at three elements often found in steel alloys. Each one affects the resulting alloy in different ways.

Manganese (Mn)

Manganese (Mn) is a powerful hardener of steel when it is combined with carbon. Manganese is also added to react with excess sulfur by producing manganese sulfide (MnS) as shown in Equation 2-7.

$$\underset{\text{(manganese)}}{Mn} + \underset{\text{(sulfur)}}{S} \longrightarrow \underset{\text{(manganese sulfide)}}{MnS}$$

Equation 2-7

If the sulfur (S) were available to combine with iron and form iron sulfide (FeS), the steel would tend to break apart during the cooling cycle because the amount of thermal expansion is different in steel and iron sulfide. Can you explain how that might lead to breakage?

Phosphorus (P)

Phosphorus (P) can increase hardness and strength of steel when it is added in large amounts, but it also tends to produce brittleness. Small amounts help to lubricate steel and increase what is called its machinability. The phosphorus (P) makes steel easier to cut, drill, or grind by reducing the size of chips that are worn away during machining. Silicon (Si) is another element that acts as a hardener, but its most important use in steel is to increase resistance to oxidation.

Chromium (Cr)

Chromium (Cr) increases the hardness of steel, its wear and abrasion resistance, and its resistance to corrosion. Chromium is added to steel in large amounts (from 11% to 20%), along with other elements such as titanium (Ti), molybdenum (Mo), copper (Cu), niobium (Nb), and selenium (Se), to make stainless steels. Stainless steels are very resistant to corrosion because a coating, called chrome oxide, forms on the surface.

Other (Mostly Nonferrous) Alloys

So far we have talked about the many kinds of steel. Many other useful alloys are made primarily from nonferrous metals such as copper (Cu), zinc (Zn), lead (Pb), tin (Sn), nickel (Ni), and gold (Au), to name but a few. Nonferrous alloys are so numerous that they cannot be described as a single group. They differ widely and have a thousand different uses. Table 2-2 shows some nonferrous alloys and their properties and uses. Keep in mind that the table is simplified. It shows only the major components of many alloys. Most alloys have more than two components, and addition of even small amounts of some elements can have important results.

Table 2-2. Nonferrous Alloys

Alloys	Properties	Applications
Copper Alloys		
High copper alloy	Melt temperature 1093°C (1999°F), good electrical conductivity	Bus bars, tubing for water, air conditioning
Copper-beryllium	Good electrical conductivity	Springs for electrical instruments
Brasses (copper and zinc)	Ability to be cast at low temperature (927°C (1701°F))	Ornamental products, jewelry, plumbing products
Bronzes (copper, tin, and other materials)	Ability to be cast, strength, corrosion resistance	Statues, plumbing products, machine parts
Copper-nickel	Corrosion resistance	Tanks and tubing for condensation, distillation, evaporation
Nickel Alloys		
Nickel and copper, chromium, and/or molybdenum	Corrosion resistance, high strength, heat resistance (depending on components)	Chemical processing equipment, food processing equipment, electrical appliances, industrial heat-resistant elements
Lead Alloys		
Lead and tin antimony	Low melting point, high ductility, high corrosion resistance	Shielding from radiation, storage batteries, caulking
Zinc Alloys		
Zinc and aluminum, copper, and/or magnesium	Low corrosion resistance, high ductility	Coatings on steel, printmaking, corrugated roofing

Activity 2-5

- Working in groups of five, discuss which alloys you would use for each application in the following list:
  - Heating element in a commercial clothes dryer
  - Radiation shielding in a nuclear medicine lab
  - Spring in an analog voltmeter
  - Drain pipe for a bathroom sink
  - Distillation tank in a chemical plant
- As a class, discuss the choices of each group. Try to arrive at a consensus for the class on each choice.
- Record your findings in your *ABC* notebook for use in the Unit Wrap-Up Activity.

Superalloys

In recent years, many new alloys have been developed. Many of them are the products of aerospace industry research. The aircraft turbojet engine was a major factor in the development of many new alloys.

The turbojet engine presented many challenges to **metallurgists** (mět′l-ûr′jĭsts)—people who work with the science and technology of metals. Metallurgists as well as chemists and materials scientists have been involved in the development of superalloys. The components in these engines are exposed to extreme conditions: high temperatures, vibration, strong centrifugal force, and corrosive gases. They must continue to perform through thousands of cycles of starting the engine, accelerating, decelerating, and stopping. This kind of cycle—repeated many times—can lead to a kind of failure called low-cycle fatigue.

The heat erosion and corrosion of these conditions would soon eat holes in carbon steels. Even the best stainless steels would have to be very thick to compensate for the continual wearing away of metal under jet engine conditions. New nickel-based materials known as superalloys have been developed to meet the challenges of turbojet conditions.

Superalloys differ from other alloys because of their microstructure. The crystals of superalloys are arranged in such a way that it is very difficult for the atoms within the crystal to be dislocated. This means that a superalloy resists deformation caused by centrifugal and shear forces that occur during flight.

Making a superalloy requires that the alloy mixture be heated and cooled in such a way that the formation of crystals is controlled. In the next section, you will find out more about how crystal formation is controlled during the making of alloys.

How Are the Properties of Alloys Controlled?

CAREER PROFILE: METALLURGICAL TECHNICIAN

Toni B. works in a company that makes forged steel products. Toni explains what is involved in **forging**.

"Forging is the process of pressing hot metal into a die by constantly applying pressure. For our main products, which are the rotating parts of jet aircraft engines, a titanium/nickel steel alloy is used.

"We start out with an ingot (a conveniently shaped mass) of steel. We melt the ingot and pour it into molds that produce smaller ingots, sized according to what we need for our product. A cylindrical ingot might be nine inches in diameter and over thirty feet long." After cooling and solidification, these new ingots are usually remelted to remove impurities from the steel. As Toni explains, for certain end products, the purity of the metal alloy is critical to the performance of the product.

The purified ingots are then pressed, extruded, or shaped by some other method into ingots that are closer to the dimensions of the final product. Then they are sawed into pieces; each piece will become a single finished product. "At this point in the process," Toni explains, "each mult—that's what we call the sawn pieces—is assigned a unique serial number. It will carry that number until completion. These numbered mults are then heated and pressed into dies to produce the rough, forged parts, called forgings."

In the quality assurance department where Toni works, various tests are carried out on the forgings. Some of the tests are called destructive tests because they destroy the mult. Tony explains, "Some of the destructive tests we perform are tensile strength tests, stress and rupture tests, and elongation tests.

"We do nondestructive testing, too," says Toni. "After the mults have gone to the machine shop for finer shaping and dimensioning, we get them for nondestructive testing. We check them ultrasonically for defects. We perform hardness tests and liquid penetration tests. Besides running the tests, we record them and note all defects in our computerized recordkeeping system. And of course, we reject defective parts."

Toni has an associate degree in chemical technology from a technical college, and she also has taken some college hours in statistics. She says that much of what she knows about metallurgy has been learned on the job.

Crystal Formation

In Subunit 1, you read that the most common form of a pure, solid substance is crystalline. Crystalline means that the atoms or molecules are arranged in fixed positions with specific distances and angles between neighboring atoms or molecules.

The crystal formation of an alloy is determined in part by the heating and cooling processes that the alloy goes through as it is made. Steel-making processes are well understood, so here we will use steel as an example.

Steel formation begins with molten iron at 2800°F (1538°C), with carbon, manganese, and silicon added. This mixture is then cooled to a solid form (the solid pieces are called ingots) in which it can be easily handled and shipped. The steel ingots are then shaped into products. Finally, the shaped products are heat-treated to give them the properties needed for specific uses.

Most steels with a carbon content above 0.30% carbon become much stronger when they are reheated to 1500°F

(815°C) and rapidly cooled by quenching in water or oil. Why does this strengthening occur?

Let's look at the cooling process in order to answer that question. As the steel cools, it begins to solidify. The crystalline structures begin to grow from a nucleus or small crystal. This crystal growth occurs at many areas in the mass of solidifying steel. As these crystals meet one another, grain boundaries form. (Each grain is a single crystal.) This grain structure repeats as the alloy cools (usually from the outside in).

Each metal has its own crystal structure. All of the grains in a particular metal have the same crystalline structure. However, as a metal cools to room temperature, the grains push against one another as they meet. The result is a distortion of the unique crystalline structure of the metal. In a metal that is cooled quickly, the grains are smaller. Smaller grains exert less pressure on one another; therefore less distortion occurs.

The smaller grains of steel that result from rapid cooling are harder and stronger than are larger-grained steels. In a smaller-grained steel, the crystals have more surface area per unit volume. (You can compare this to the composition of concrete. Concrete made with sand and gravel is stronger than concrete made with larger rocks.)

Another critical factor in the strength of steel, besides the size of the crystals or grains, is the shape of the crystals. Depending on how steel is processed, different structures maybe formed. One crystalline structure may be transformed into another as a result of heat treatment. Table 2-3 shows four basic crystalline structures found in alloys.

Table 2-3. Four Crystal Structures of Alloys

Crystalline form name	**Crystalline structure**	**Metals with form**	**Steel form and transition temperature**
Body-centered cubic (BCC)	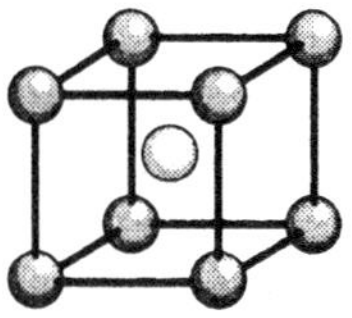	Fe, Cr, Mo, Ta, W, V, Nb	*Ferrite form* The ferrite form develops when the steel is cooled very slowly. It forms in the temperature range from 590°C to 705°C. This form of steel is soft and very ductile.
Face-centered cubic (FCC)	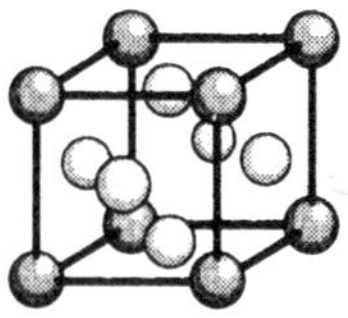	Fe, Al, Cu, Au, Pb, Ni, Pt, Ag	*Austenitic form* The austenitic form of steel develops when the steel is heated to above 760°C. This form of steel exists only at elevated temperatures.
Body-centered tetragonal (BCT)	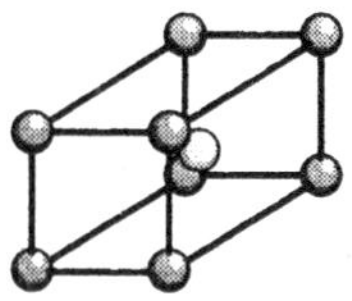	Fe	*Martensitic form* The martensitic form of steel develops when the steel is cooled quickly. It forms in the temperature range from 102°C to 290°C. This form of steel is hard.
Hexagonal close packed (HCP)	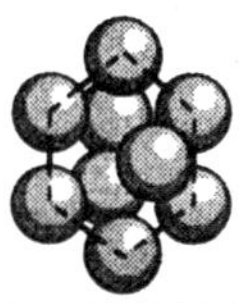	Cd, Co, Mg, Ti, Zn	

Activity 2-6

- Form the class into four groups. In the groups, decide if each item in the list below should be hard, soft, or intermediate. After determining the hardness needed, decide if the item should be cooled quickly or slowly to get the desired hardness.
  - Head of a sledgehammer
  - Bolts attaching to the cylinder head of an engine block
  - Teeth on the digging edge of a back hoe
  - Blade on a pocket knife
  - Pin holding the two halves of a hinge together
- After each group has determined the hardness requirements and cooling rate, discuss as a class the decisions for each item in the list and arrive at a consensus for each item.

Looking Back

Metals have certain properties in common: they are good conductors of electricity and heat; they are ductile and malleable and can be polished to a sheen; they can be oxidized. The specific properties of metals are related to their atomic structure, their bonding characteristics, and their crystalline structure. The atomic structure of metals accounts for their reactivity. Metallic bonding is responsible for physical properties such as hardness and tensile strength.

The way steel and other alloys are processed has a direct effect on their final properties. The reheating and cooling processes are used to control the crystal formation and grain size of steel products, resulting in the desired hardness and strength.

Further Discussion

- A drill bit on a power drill must be hard enough to cut other metals. If a drill bit is improperly used at excessive speeds and becomes hot, how will the properties of the steel and the usefulness of the bit be changed?

- When metals are welded, they are heated to a molten state. How might welding affect a heat-treated, hardened steel?

Activities by Occupational Area

General

Precious Metals

- Select a precious metal—gold, silver, or platinum—to research. Find out about its history and its current uses. In your *ABC* notebook, write a brief report.

Galvanized Metals

- Talk to a person who sells metals used for building materials. Find out what galvanized metal is and what materials are usually galvanized. What other methods are used to prevent oxidation of metals?

Agriscience

Agricultural Mechanic

- Contact Future Farmers of America and find out what an agricultural mechanic does and by whom such a person is employed. What kind of machining and/or welding skills are required for someone in this field? Use your findings to write in your *ABC* notebook a job description for an agricultural mechanic. (Your teacher will provide you with a sample job description for another type of job.)

Health Occupations

Biomedical Technicians and Biocompatibility

- Contact a biomedical equipment technician and find out what considerations these technicians must have for the principle of biocompatability. Report your findings to the class.

Family and Consumer Science

Properties of Cutlery

- At a cookware or hardware store, look at the different types of cutlery (knives) used for cooking. Find out from a salesperson or from available product information the kinds of metals from which they are made. Make a chart of the different available brands, the metals from which they are made, and their advantageous and disadvantageous properties. Consider ease of use, life of the product, care requirements, and so forth.

Industrial Technology

Corrosion of Pipe Joints

- When certain dissimilar metals are joined in a plumbing system, corrosion can occur rapidly. From your teacher, get a sample of each of the following metals: magnesium ribbon, aluminum foil, steel wool, and copper. Get a 600-ml beaker and fill it with about 400 ml of distilled water. Connect the copper and aluminum by bending the aluminum foil securely around the copper.
- Also connect the steel wool and the magnesium ribbon by twisting the steel wool and one end of the magnesium ribbon together. Place the two attached pairs of metal samples in the beaker filled with distilled water. Allow the samples to remain in the water for 24 hours or more. After the 24-hour period, observe the metals and the distilled water. Record your observations in your *ABC* notebook. Present your findings about these metals to the class. Use the metals and the distilled water as visual aids for the presentation.

LAB 3

HOW IS THE HARDNESS OF METALS MEASURED?

PREVIEW

Introduction

Beverly R. works as a purchasing assistant in a machine shop. Her supervisor has directed her to check out a potential problem. A supplier of round steel-bar stock has delivered a batch of material that is not color coded. The color code is important because it indicates that the steel has certain chemical properties. The steel might also have physical properties, such as hardness, that could damage cutting tools. Many different grades of steel, each with different chemical and physical properties, are used to manufacture products requiring those properties.

Beverly goes to the storage area and locates the batch of material. She rubs a file across several of the steel bars. The file "bites" into the steel and, as a result, Beverly knows the material is soft enough to be cut with a saw. She then asks the shop foreman to cut an inch-long piece from the end of one of the bars. The foreman gives this piece to the shop's metallurgist, who will analyze its chemical properties.

Purpose

In this lab, you will test the hardness of various metals using two methods.

Lab Objectives

When you've finished this lab, you will be able to—

- Compare two different methods for testing the hardness of a metal.
- Rank a series of metals in order of their hardness.

Lab Skills

You will use these skills to complete this lab—

- Use a file to scratch metals.
- Use a ruler to measure diameter.
- Use a magnifying lens to aid in taking measurements.

Materials and Equipment Needed

metal samples
file card
marking crayon
hand lens, or stereoscope (10X)
flat file
hardness testing apparatus
machinist's rule

Pre-Lab Discussion

The hardness of a material is the degree of resistance it shows to penetration. Metals range in hardness from relatively soft metals such as lead to relatively hard metals such as titanium. The harder a metal, the better it will resist wear. For some metals, hardness is an indication of other mechanical properties, such as machinability and tensile strength. Many metals, especially steel, can be increased in their hardness by heating them and carefully controlling the rate at which they are cooled.

The hardness of metal products is routinely tested during their manufacture using standards set by the American Society of Testing and Materials (ASTM). The results of such tests help tell whether a material will be suitable for a given application. Hardness is also used by metallurgists when evaluating failures of metal products. They can check whether the correct heat treatment was performed, or if an unauthorized heating and cooling cycle was used during fabrication.

The hardness of metals is determined by a number of methods. The introduction to this lab referred to a qualitative way of estimating hardness—the file-hardness test (Figure L3-1). Hardness tests on metals measure the resistance of a metal sample to penetration. Machines provide accurate measurements of hardness by applying a specified amount of force to a diamond tip or a hard steel or carbide ball, causing it to penetrate the surface of the sample (Figure L3-2). Either the depth of penetration of the ball or the diameter of the indentation it made on the surface of the sample is then measured. This measurement is located in a table of hardness values for the type of test performed.

In this lab you will test the hardness of metal samples by indenting them with a hard steel ball that is attached to the bottom of a falling weight. You will then measure the diameters of the indentations and use them to compare the hardness of the samples.

Safety Precautions

- Be careful not to cut your skin when using the file.
- Keep your hands away from the base of the hardness testing apparatus when your lab partner is dropping the weight.

LAB PROCEDURE

Method

Work in groups of three or four.

Part A. Testing File Hardness of a Metal

1. Obtain a set of metal samples whose hardness you will test. Using a marking crayon, number them 1, 2, 3, and so forth, ½" from the edge of the samples.

2. Applying a strong, steady pressure, drag the file along the first sample as shown in Figure L3-1. Observe the result and describe this result in the data table.

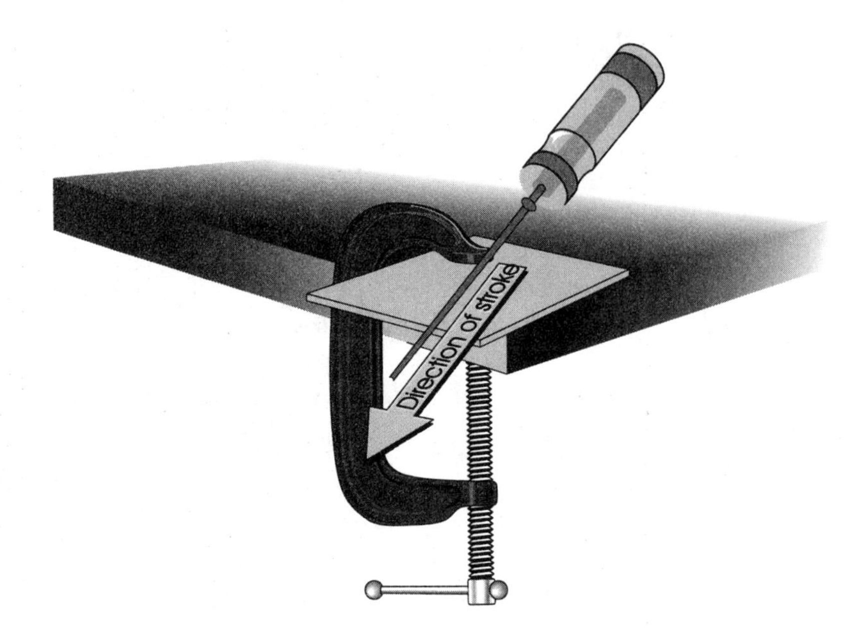

Figure L3-1 Testing file hardness of a metal

3. Repeat Steps A1-2 for each of the other metal samples. Clean the file with the file card after each use. Drag the file along the samples with the same amount of pressure each time.
4. Answer Conclusion Question 1.

Part B. Testing Hardness of a Metal Under Load

5. Obtain a sample of metal from Part A whose hardness you will test. Write the number of the sample in the data table. Check both surfaces of the sample for indentations. Using a marking crayon, color inside these indentations so that you can distinguish them from the new indentation you will make in the sample.

6. Place the hardness tester on an uncarpeted floor. Remove the dropweight from the guide rods (or guide tube) and place it on the floor or on the lab counter.
7. Place the metal sample on the base of the hardness tester. Align the sample beneath the guide rods (or guide tube) so that your indentation will be made at least ½" from the edge of the sample (see Figure L3-2).

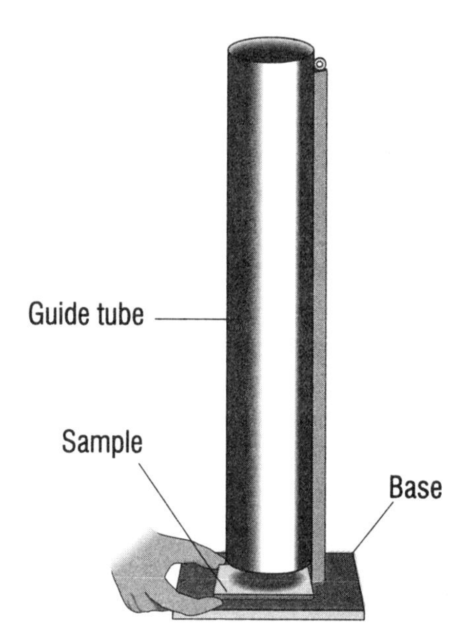

Figure L3-2
An apparatus for measuring hardness

8. Make sure that your lab partner does not have his/her fingers near the base of the hardness tester. Then insert the dropweight (with the indenter pointed downward) into the guide rods or tube and let it fall onto the metal sample.
9. Remove the weight from the guide rods or tube and place it on the floor or lab counter. Remove the metal sample from the base of the tester.
10. Find the indentation made in the sample by the dropweight.
    - If the indentation is at least one diameter from the nearest indentation, measure the largest diameter of this indentation to the nearest $1/64$-inch, using your machinist's rule and a magnifying lens as shown in Figure L3-3. Record this diameter in the data table. Share your data with the class.
    - If the indentation is within one diameter of any other indentation do not measure it. Mark it with the marking crayon and repeat Steps B7-10.

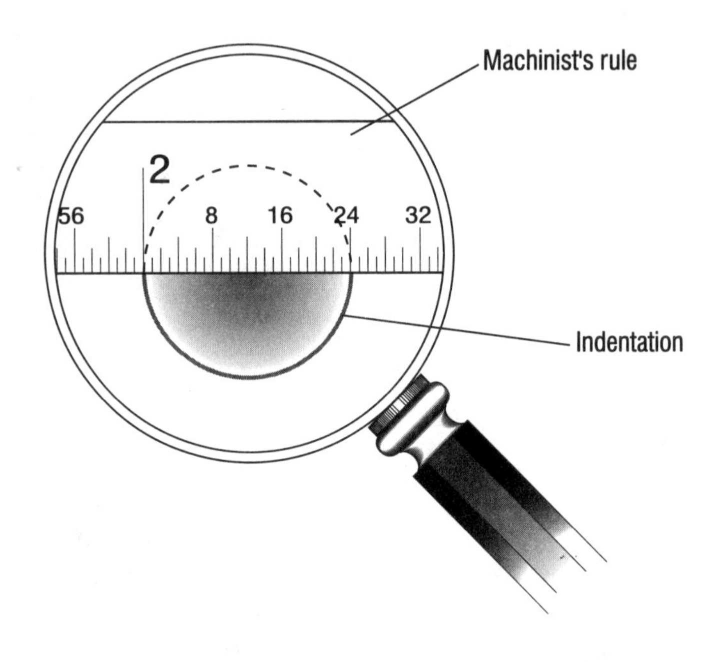

Figure L3-3
Measuring the indentation made by the dropweight

11. Repeat Steps B5-10 for each metal sample. Give the results of your measurements to your teacher, who will include them in the class data table.

Cleanup Instructions

- Clean up any metal shavings from your lab table.
- Return the metal samples to the lab counter.
- Remove the dropweight from the hardness tester.
- If you used a stereoscope in this lab, return it to the proper storage cabinet.

Observations and Data Collection

Data Table

Type of Metal	Result of File Hardness Test (Part A)	Diameter of Indentation (Part B)

Calculation

A. Compute the average diameter of all indentations made by the class for each metal. Use the formula:

$$\text{Average diameter of indentation in metal sample} = \frac{\text{Sum of diameters in all indentations}}{\text{Number of indentations}}$$

Round your final answer to the nearest $\frac{1}{64}$th inch and write it in your *ABC* notebook.

WRAP-UP

Conclusions

1. Based on the results in Part A, rank the metal samples tested by your group from most hard to least hard. Write this ranking in your *ABC* notebook. Discuss your results as a class and come to an agreement on how the metals tested by the class should be ranked.
2. Observe closely the shape of the indenter on the dropweight. In your *ABC* notebook make a drawing of a cross section of the indenter after it has penetrated the sample. Use this drawing to demonstrate to the class how hardness of the metal sample is related to the diameter of the indentation.
3. Based on the results in Part B, rank the metals in order from most hard to least hard. How does this order differ from that for the same metals tested in Part A?
4. Which of the two hardness tests used in this lab do you prefer? Explain.

Challenge

5. Do you think your hardness tester could be used to measure any of the following nonmetal materials:
   - a piece of slate
   - a piece of ceramic tile
   - a board
   - a rubber mat

   Explain your answer for each of the materials, then suggest a way you might be able to measure its hardness.

LAB 4

CORROSION OF METALS

PREVIEW

Introduction

Paul D. is a plumbing contractor. We called him one day recently to ask what causes metal corrosion in plumbing materials. His answer centered on two main issues: 1) electrolysis and 2) the use of liquid chemicals to clear drains.

Paul explained that electrolysis occurs in plumbing fixtures when two dissimilar metals are joined, with water running through them. "If I were to connect two unlike metals, such as copper and steel," he told us, "I would be setting up the same conditions you find in a battery. The steel would begin to give up ions and, as a result, it would eventually corrode." To prevent electrolysis from occurring, Paul connects dissimilar metals by using a nonconductive coupling, usually a pipe fitting made of nylon.

"Another cause of corrosion is the use of some of the liquid plumbing products," says Paul. "These used to be sodium hydroxide; now they also generally contain sodium hypochlorite. Most garbage disposals have a metal grinding case that is made of a cheap alloy, and it is easily corroded by the 'plumbing chemicals' that people purchase in the supermarket."

Paul says that his knowledge of chemistry, especially metallurgy, and biology, helps him in his job as a plumber. He has a high school diploma and attended college for one year in computer science. To become a plumber, he was required to work as an apprentice for three years, to pass the journeyman plumber test, to work for an additional year, and to pass the master plumber test. The master plumber certificate is required of people who work as independent plumbing contractors.

Purpose

In this lab, you will observe the electrical potential between two metals immersed in an electrolyte solution.

Lab Objectives

When you've finished this lab, you will be able to—

- Identify chemical change on a metal sample.
- Predict the outcome of connecting two metal samples.

Lab Skills

You will use these skills to complete this lab—

- Measure voltage with a digital voltmeter.
- Make electrical connections between two metal samples.
- Observe metal samples for evidence of chemical change.

Materials and Equipment Needed

graduated cylinder
3% sodium chloride (NaCl) solution
3-inch piece of magnesium (Mg) ribbon
1-inch × 3-inch piece of aluminum (Al) foil
1-inch × 3-inch piece of copper (Cu) foil
steel wool
600-ml beaker
2 copper wire leads with alligator clips on each end
digital volt meter with a 2-volt full-scale setting
lab apron
safety goggles

Pre-Lab Discussion

Some metals oxidize (lose electrons) more easily than others. The more easily oxidized metals can protect other metals from oxidation when they are exposed to oxidizing agents in their surroundings—for example, the zinc used to protect the iron in the bridge reinforcements that were discussed in the text.

There are also some situations where joining dissimilar metals can cause problems. For example, if copper and aluminum pipes are joined in a plumbing system, one of the metals will protect the other metal from corrosion. However, the joint will begin to leak as the "protecting" metal oxidizes.

When two different metals are immersed in an electrolyte solution like sodium chloride, an electrical potential is set up between the two metals. If the metals are connected by an electrical conductor, current will flow and oxidation will take place at the negatively charged metal. A voltmeter will measure the electrical potential (voltage) between the two metals and tell you which metal is positive (being reduced) and which metal is negative (being oxidized). These data can be organized into a list that is ordered from more easily oxidized to less easily oxidized. This list is called an electrochemical series.

Safety Precautions

- Magnesium metal reacts with water. Use only magnesium ribbon. Do not use powdered magnesium.

LAB PROCEDURE

Method

Put on your lab apron and goggles.

1. Pour 400 ml of 3% NaCl solution into a 600-ml beaker.
2. Place the four metals in the beaker so that one end is under the NaCl solution and the other end is above the surface as shown in Figure L4-1.

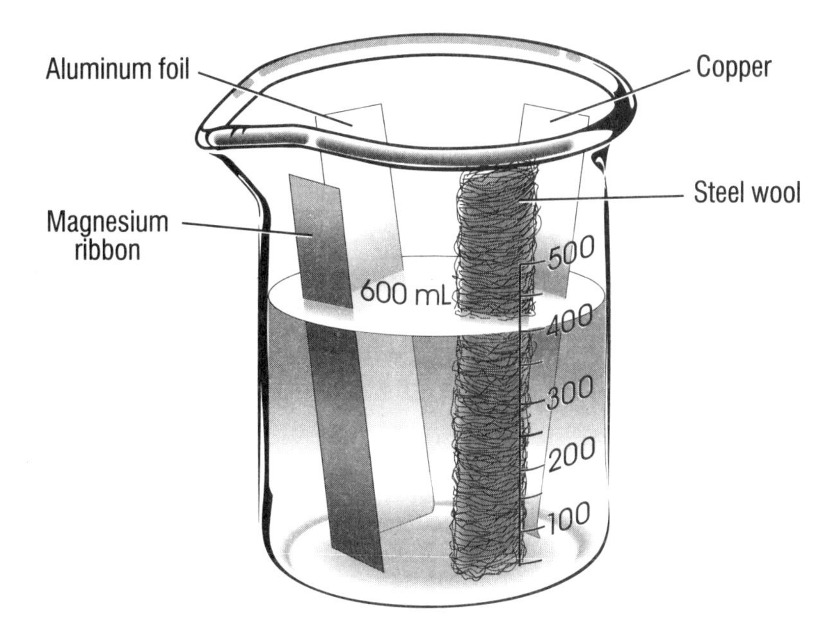

Figure L4-1 Arrangement of metals in beaker

3. Measure the electrical potential between each of the different metals with the digital voltmeter set at 2 volts. Touch the common (black) lead to the magnesium ribbon and touch the positive (red) lead to each of the other metals. Allow time for the meter reading to stabilize and record the reading for each metal pair in Data Table 1. Then, move the common lead to the aluminum foil and read the voltage with the copper and iron samples. Record these readings in Data Table 1. Finally, move the common lead to the steel wool and measure the voltage with steel wool and copper.

Data Table 1

	Mg	**Al**	**Fe**	**Cu**
Mg	0.0 V			
Al		0.0 V		
Fe			0.0 V	
Cu				0.0 V

4. Use the wire leads and connect the magnesium ribbon to the steel wool and the aluminum foil to the copper. Observe the metals for 25 minutes. Record your observations in Data Table 2.

Data Table 2

Metal	Observation
Mg	
Al	
Fe	
Cu	

Cleanup Instructions

- Place the metals in the containers provided by your teacher.
- Empty, wash, and dry the beakers. Return the cleaned beakers to their proper storage location.
- Return the copper leads and voltmeter to their proper storage locations.

Calculation

A. Based on your data for this lab, order the metals aluminum, copper, iron, and magnesium from most reactive to least reactive under these conditions.

WRAP-UP

Conclusions

1. Based on your data, predict what you would observe if a piece of copper and a piece of magnesium were placed into a 3% NaCl solution and connected with a copper wire.
2. Based on your data, predict what you would observe if a piece of steel wool and a piece of aluminum foil were placed into a 3% NaCl solution and connected with a copper wire.

Challenge

3. Suggest how you could use magnesium or zinc (zinc is slightly less reactive than magnesium) to protect the iron in a drilling platform used to drill offshore oil wells.

SUBUNIT 3

CERAMIC MATERIALS

THINK ABOUT IT

- Why are metals used instead of ceramics to build automobile bodies?
- Why are most ceramics insulators?
- How did the revolution in electronics become possible?
- How far can electronics miniaturization go?

SUBUNIT OBJECTIVES

After you complete this subunit, you will be able to—

1. Distinguish ceramics from other materials such as metals and polymers on the basis of their chemical structure and properties.
2. Relate the differences in structure between crystalline and glass ceramics to the differences in their properties.
3. Investigate a selected ceramics application in modern technology.
4. Use sol-gel synthesis to make a simple ceramic material.
5. Explain why high temperatures and atmospheric control are necessary for ceramic manufacturing processes.
6. Compare the accuracy of two methods for determining the volume of an object.
7. Determine the density of a liquid.
8. Recommend how porous bricks should be for a given application.
9. Perform a redox titration to determine the oxidation state of an element with variable oxidation states.

PROCESS SKILLS

You will use these skills in lab—

- Measure the weights of solids and liquids using a balance.
- Measure the volume of a liquid using a graduated cylinder.
- Measure the volume of a liquid with a pipette.
- Titrate a reaction mixture to a visually determined endpoint.

What Are Ceramics?

Ceramics are chemical compounds that are made generally by firing or processing **raw materials** at high temperatures. Ceramics generally have high melting points, are heat resistant, and are poor conductors of electricity.

Ceramics were among the first materials used by people. These first ceramics were in the form of native stone or bone tools, rock, and clay. Bone consists of calcium phosphate. Rock, stone, and clay are mostly silicon and oxygen. Silicon and oxygen are abundant and together account for almost 75% of the material in Earth's crust, as shown in Figure 3-1.

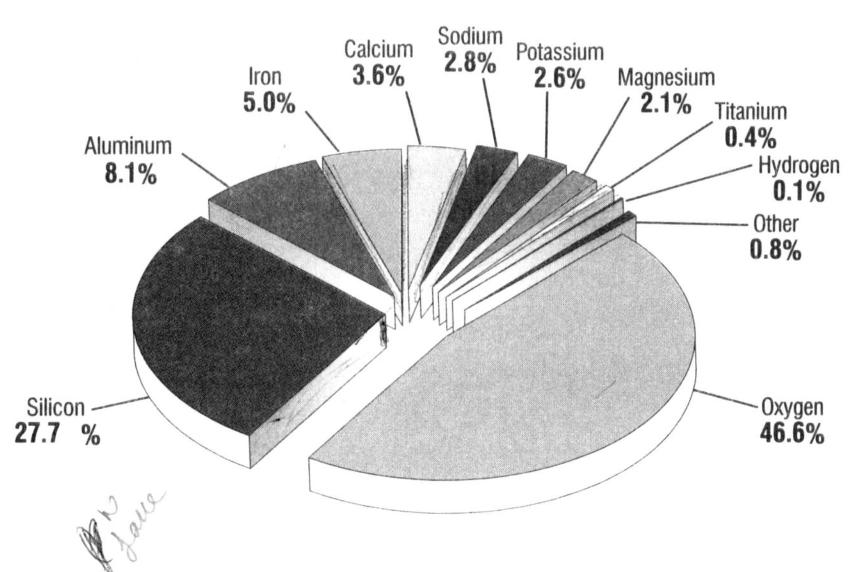

Figure 3-1 Abundance of elements in Earth's crust

Originally, people used these materials in the natural state for stone weapons and tools. Slowly, people learned to improve upon naturally occurring ceramics. For example, pottery is burnt clayware and was used as far back as 4000 B.C. Other structural clay products such as brick, tile, sewer pipe, sanitaryware, and dinnerware were developed later.

CAREER PROFILE: SALES ENGINEER

David C. is a sales engineer for a ceramic materials supply company. The company supplies fundamental ceramics materials to manufacturers that use ceramics in their products. Such products include traditional ceramic materials such as tiles, toilets, and bricks and ceramics used in the electrical power distribution and electronics industries.

Much of David's job involves relaying information between suppliers—the companies that produce ceramic powders and chemicals—and manufacturers—the companies that make products out of ceramic materials. He gives an example of how this works. "I have a number of customers," he begins, "who purchase frits through our company. A frit is a type of glass ceramic into which you can put other materials that are less stable or more difficult to work with. For example, a material called barium oxide is used in some ceramic materials, but it can introduce health hazards to workers unless it is put into a frit."

David continues, "I've been able to locate a very high-quality, low-priced frit made in Mexico. The clays in this area of Mexico are purer than most clays available in the United States, so the supplier there is able to make a purer frit than other producers can. I relay to the Mexican producer the formula for the frit that is required by my customers."

When asked why manufacturers don't find ceramics producers themselves, David explains that keeping up with the supplier market is a very large task, one for which most manufacturers do not have time. When he looks for producers, David has to assess the quality of their product and consider whether they will be able to provide a steady supply of the material needed at a cost affordable to his customers.

"Sometimes a manufacturer will call me and say, 'we're thinking of making such-and-such a product, but first we need to find out if the materials are available at a reasonable cost.' I go to the suppliers to get the needed information."

David has a master's degree in ceramics engineering and describes his position in sales as somewhat unusual. "Most ceramics engineers work for ceramics producers or ceramics product manufacturers," he says. "But there is a need for someone with my background in sales, I have to know the ceramics language and understand the manufacturing processes. I have to be able to work out formulas and run tests on materials. In fact, I spend some of my time in the laboratory, conducting flow tests on glazes, viscosity tests, and checking manufacturers' formulas."

In recent years, scientists and engineers have learned to make new types of ceramics that offer potentially useful properties. The development of many of these ceramics has revolutionized modern technology. Superfast computers, cordless portable devices, fiber-optic communication networks, dental implants, and hip-joint replacements are a few examples of the influence of ceramics in the modern world.

Traditional and Advanced Ceramics

Ceramics that have been used for many years for products like pottery, brick, tile, sanitaryware, dinnerware, porcelain, concrete, abrasives, refractories, and glass are called traditional ceramics (see Figure 3-2).

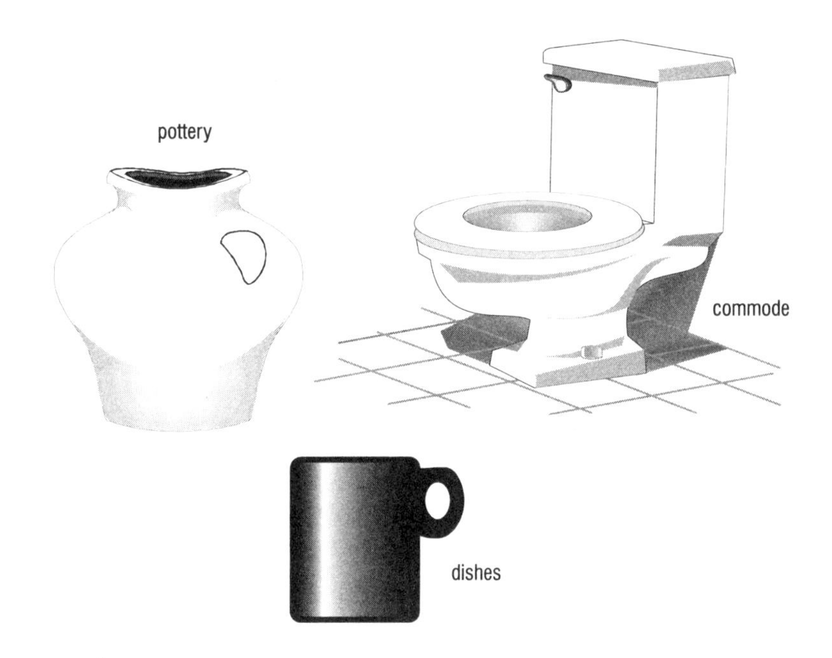

Figure 3-2 Traditional ceramics

Many of these come from naturally occurring clay materials. Recent developments in ceramic materials have caused a new term to be coined—advanced ceramics (called fine ceramics in Japan). Advancement in this area of materials science allows engineers to design a ceramic material with exactly the properties needed for a particular use. Some of the new uses of ceramics are found in the aerospace, automotive, and electronic industries. The properties of materials are controlled by making the materials according to a specified chemical composition and internal microstructure. Breakthroughs in synthesis and processing procedures have made this possible.

Traditional ceramics come primarily from natural raw materials formed by simple processes. Advanced ceramics often use synthetic raw materials and complex processes.

Traditional ceramics have simple applications. Advanced ceramics often have very demanding requirements. For example, some advanced ceramics used in spacecraft have to withstand extremely high and extremely low temperatures. (Temperatures during takeoff and reentry are very hot, but temperatures in outer space are very cold!)

The properties of traditional ceramics can be evaluated with simple instruments. Advanced ceramics require complex instruments to analyze the relationship between the properties and the arrangement of atoms in a ceramic material.

What Materials Are in Ceramics?

Ceramic materials are chemical compounds made of metallic and nonmetallic elements. For example, the combination of aluminum, a metal, with oxygen, a nonmetal, gives the ceramic aluminum oxide. This reaction is shown in Equation 3-1.

$$\underset{\text{(aluminum)}}{4Al(s)} + \underset{\text{(oxygen)}}{3\,O_2(g)} \longrightarrow \underset{\text{(aluminum oxide)}}{2\,Al_2O_3(S)}$$

Equation 3-1

Figure 3-3 shows the periodic table. The symbols for the metallic elements in this table are shaded. Compounds of any of these metallic elements with any of the unshaded nonmetallic elements, like carbon, nitrogen, oxygen, fluorine, phosphorus, or sulfur, can produce a ceramic material.

Sometimes combinations of nonmetallic elements can produce ceramic materials also. For example, both ice (H_2O) and silicon carbide (SiC) are ceramic compounds made only of nonmetallic elements. These nonmetallic compounds are classified as ceramics because their properties are similar to the general properties of other ceramics.

The term ceramic comes from a Greek word that means "burnt stuff." This is an appropriate term because many of the desirable properties of ceramics are achieved only through a high-temperature process called firing.

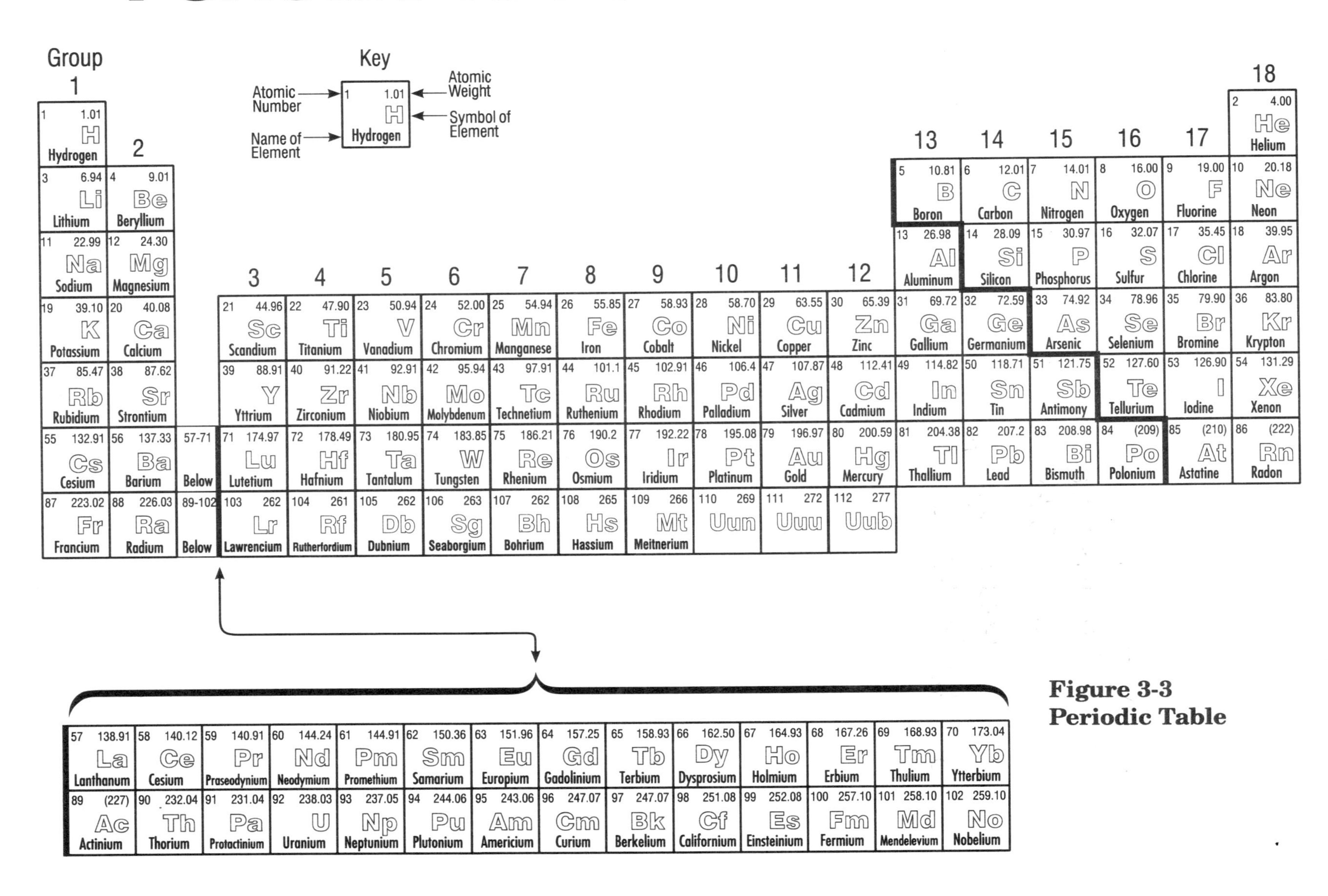

Figure 3-3
Periodic Table

Ceramics Are Not Always Simple Compounds!

Ceramics are not always simple compounds made of a single metal and a single nonmetal or two nonmetals. Some ceramics are as simple as silicon carbide. However, they can be very complex like the **superconductor**, $Tl_2Ba_2Ca_2Cu_3O_{10}$, a compound with four different metals and the nonmetal oxygen. Some of the most useful ceramics not only have complex formulas, but the ratio of atoms can vary to give slightly different properties. SIALON, a strong, hard engineering ceramic made of silicon, aluminum, oxygen, and nitrogen, is an example. The general formula for SIALON is $Si_{100-x}Al_xO_{200-x}N_x$, where x represents the variable percentage (by atomic composition) of silicon replaced by aluminum in the ceramic.

Activity 3-1

- A sample of SIALON has 12% aluminum atoms replacing silicon. How would you write the formula for this SIALON ceramic?
- Write formulas for the following aluminum percentages of SIALON—8%, 10%, 14%, 16%
- Compare your formula to the formulas of others in your class. Discuss any differences and agree on a single formula.
- As a class, discuss the name SIALON. Where does this ceramic get its name?

What Kinds of Bonding Occur in Ceramics?

In ceramic materials, the main types of bonding are ionic and covalent.

Ionic and Covalent Bonding

Ionic bonding is the bonding between atoms that have a large difference in electronegativity. Electronegativity is the tendency for an atom to accept other electrons. In an ionic bond, the atom with the higher electronegativity gains an electron and becomes an anion (negative ion). The atom with the lower electronegativity loses an electron and becomes a cation (positive ion).

In the case of ionic bonding, anions and cations in a crystal occupy specific positions. The anion's negative charge is attracted to the positive charges of all the cations that surround it. At the same time, the anion's negative charge is repelled by the negative charges of all the anions around it. Therefore, in a crystal, the anion will be closer to the surrounding cations than the surrounding anions. The same is also true for the cations—that is, they will be closer to the surrounding anions in the crystal then they will be to the surrounding cations. This can be seen in Figure 3-4. The larger spheres represent the anion or chloride ions, and the smaller spheres represent the cations or sodium ions. Notice that the closest neighbor of every chloride ion is a sodium ion (the closest neighbors are the closest atoms that are on any of the three straight lines that run through each atom). Also, notice that the closest neighbors of every sodium ion are chloride ions.

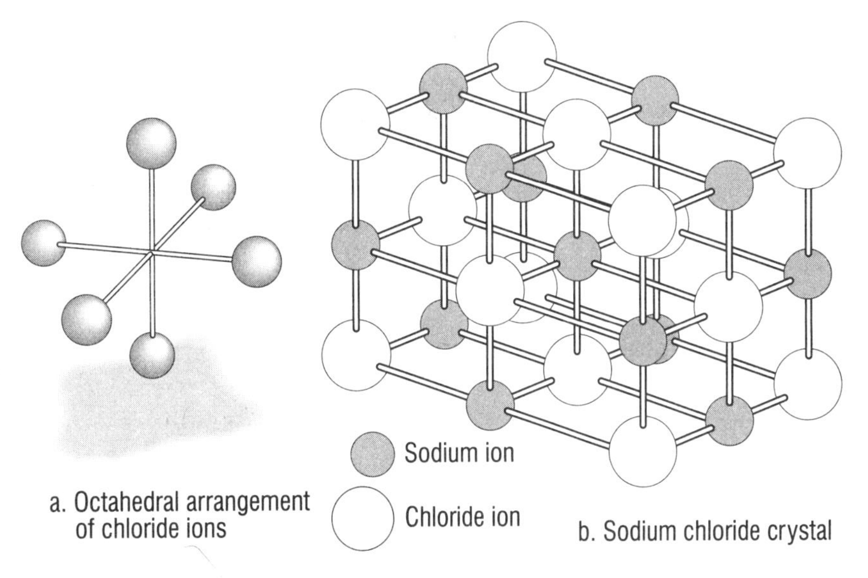

Figure 3-4 Sodium chloride crystal structure

In ionic bonding, the ~~elements~~ are localized around the anions. Since the electrons are localized, materials that are bonded by ionic bonding do not conduct electricity in the solid state. The attraction of opposite charges and the repulsion of like charges keep both the anions and cations locked into the lattice positions. This causes the materials to be brittle if they are hard or to crumble if they are soft. Ionic materials are not malleable or ductile.

What is covalent bonding? Covalent bonding is the bonding between two nonmetal atoms of approximately equal electronegativity. The electrons are shared equally between the two atoms. In covalent bonding, the electrons are localized between the two atoms. Covalently bonded materials do not conduct electricity. However, depending on the nature of the molecules, these materials may be ductile or malleable under certain conditions.

Percent Ionic Character

Most of the time, the bonding in ceramics is a combination of ionic and covalent bonding. The percentage of ionic character in a bond can be calculated based on the electronegativities of the bonded atoms. Table 3-1 lists the percent ionic character for several compounds along with the electronegativities of the bonded elements.

Table 3-1. Ionic Character

Compound	First Element	Electro-negativity of First Element	Second Element	Electro-negativity of Second Element	Electro-negativity Difference	Percent Ionic Character
Cesium fluoride	Cesium	0.82	Fluorine	4.00	–3.18	92.02
Potassium fluoride	Potassium	0.84	Fluorine	4.00	–3.16	91.76
Sodium chloride	Sodium	0.96	Chlorine	3.00	–2.04	64.67
Aluminum oxide	Aluminum	1.63	Oxygen	3.37	–1.74	53.09
Silicon nitride	Silicon	1.94	Nitrogen	2.81	–0.87	17.24
Silicon carbide	Silicon	1.94	Carbon	2.56	–0.62	9.16
Diamond	Carbon	2.56	Carbon	2.56	0.00	0.00

What do you find out when you determine the ionic character of a compound? The ionic character tells you something important about the coulombic attractive forces between the atoms in the compound. Coulombic forces are the attractive forces between positive and negative charges. A compound like silicon nitride is mostly covalent, but has a 17.24% ionic character. This means that it is slightly **polar** (has a separation of charges from one end of the molecule to the other). Its slight polarity means that the coulombic attractive forces between silicon and nitrogen atoms are a little stronger than those between the carbon atoms in diamond, for example. Diamond has a pure covalent bond. How would you compare that with the bonding in potassium fluoride?

Crystalline and Noncrystalline Ceramics

The atoms in ceramics are arranged in a particular manner. Their arrangement—which is called their **microstructure**—can take one of two forms—crystalline or **amorphous** (ə-môr′fəs) (noncrystalline). These two microstructures give a material very different properties. For some uses of ceramics, a crystalline microstructure may be needed. For others, an amorphous arrangement may be needed.

Activity 3-2

- Get one hundred marbles and a clear plastic box from your teacher. Each marble represents an atom in a crystal. Place one layer of marbles in the plastic box. Call this layer a. Arrange the marbles so you can get the most marbles in this layer. Locate the triangular shaped empty spaces labeled B and C in Figure 3-4.
- Add a second layer on top of the first layer. Call this layer b. Can you cover both the B and C empty spaces with a single layer?
- Add a third layer on top of the second layer. If you cover the empty spaces in layer a that were not covered by the second layer, label the third layer c. This arrangement is called "...abcabcabc...." If you do not cover the empty spaces that were not covered by the second layer, the arrangement is an "...abababa...."
- "In a class discussion, design a method of measuring the space occupied by the marbles and the empty space in each arrangement. Possible equipment to use is a graduated cylinder and water. As an *ABC* project, determine which arrangement,"...abcabcab..." or "...abababa..." is the most efficient at filling the space.

The atoms in ceramics are arranged in a specific manner. In the sodium chloride crystal, the chloride ions form a close-packed arrangement with an "...abcabcab..." layer sequence. This arrangement of layers is called cubic-close packing or CCP (see Figure 3-5). When the layers are in the "...abababa..." sequence, the arrangement is called hexagonal-close packing or HCP. One type of empty space in the CCP arrangement is surrounded by six spheres all an equal distance from the center of the empty space in an octahedral arrangement as shown in Figure 3-6a. In sodium chloride, this empty space is occupied by a sodium ion. This gives the sodium chloride crystal the structure shown in Figure 3-6b. This regular arrangement of atoms is repeated throughout the crystal, giving the sodium chloride crystal what is called long-range order. Such a material is called a crystalline ceramic.

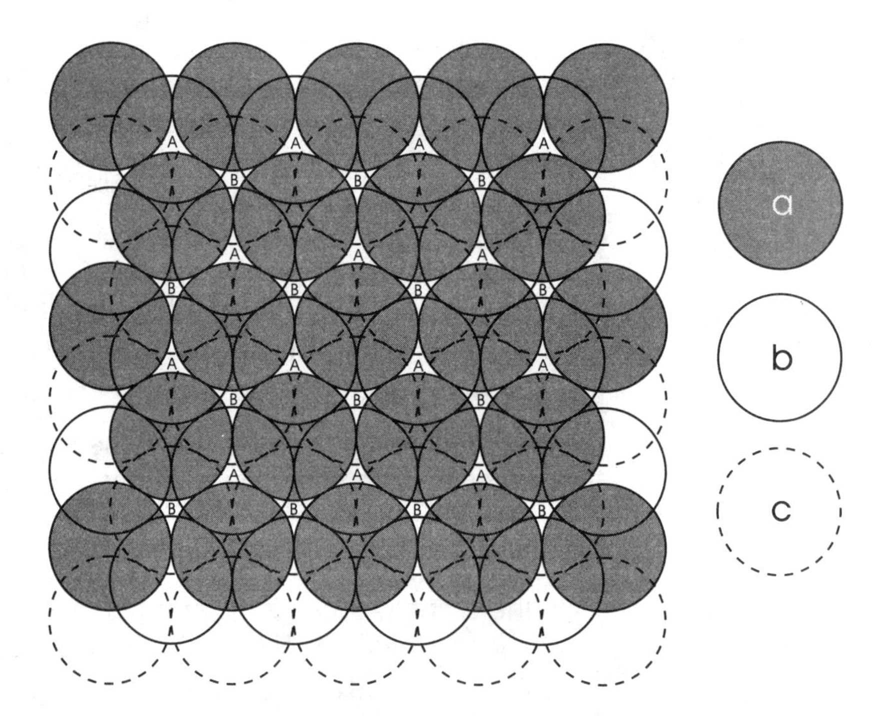

Figure 3-5 Layers of atoms in cubic-close packed arrangement

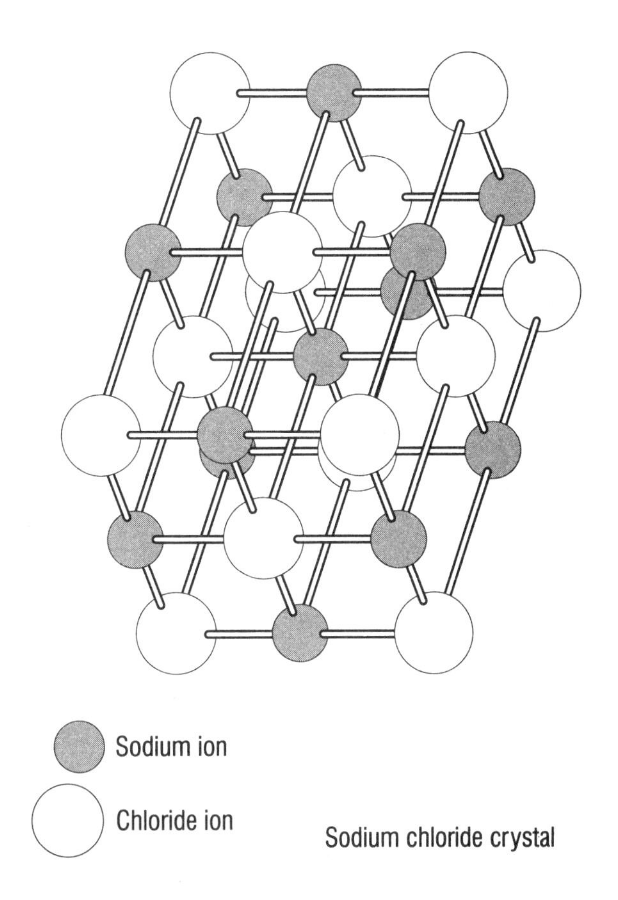

Figure 3-6 Geometry of sodium chloride crystal

In some ceramics, the regular arrangement of atoms may last over only a short distance. These ceramics may not have a long-range repeating structure. Such ceramics are called noncrystalline or amorphous ceramics. For example, the difference between crystalline and noncrystalline SiO_2 is shown in Figure 3-7. In crystalline SiO_2 (Figure 3-7a), a regular, repetitive, long-range order exists throughout the material. In noncrystalline SiO_2 (Figure 3-7b), each Si atom is surrounded by the same number of oxygen atoms, but variations in the O-Si-O bond angles cause a loss of the long-range order. Glass is another term for this type of amorphous or noncrystalline ceramic material. Transparent window glass is this form of SiO_2.

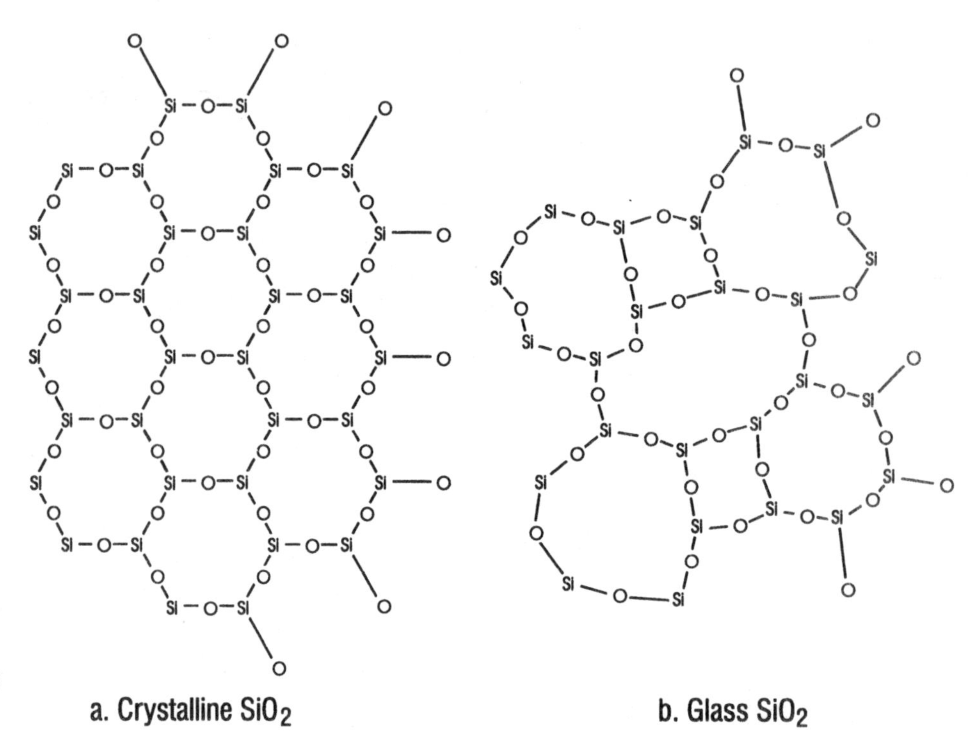

Figure 3-7 Representation of crystalline and glass forms of SiO_2

How Do Crystalline and Glass Materials Differ in Formation?

Crystalline and glass materials differ in their formation. Figure 3-8 is a graph representing the change in state from a solid to a liquid for two materials during cooling and changing from a liquid to a solid. After the glass cools, its density (mass per unit volume) increases gradually. The **molten** glass becomes more and more viscous (slow flowing) until it finally becomes a solid. But something very different happens to the crystalline material as it cools. When it reaches a certain temperature, it suddenly solidifies and its density increases greatly.

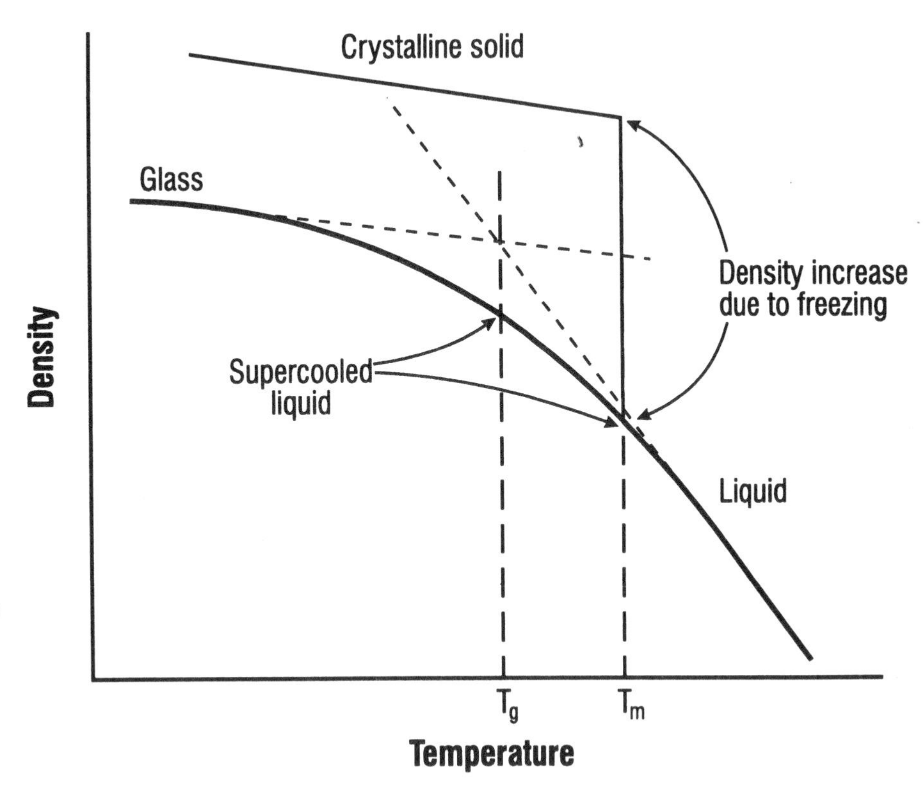

Figure 3-8
Graph of density versus temperature for crystalline and glass material

The people who make crystalline and glass materials carefully control every aspect of the heating and cooling process in order to achieve a material with the microstructure they want.

Activity 3-3

- Visit a glass studio or contact a glass manufacturing company and arrange to tour the plant.
- On your visit, collect data that will help you to develop a flowchart of the glass-making process. You can use simple boxes or circles to show different stages in the process, or you can use slightly more artistic drawings, as you wish. The important thing is to show what happens to the glass from start to finish. Be sure to include the temperatures of furnaces and the temperature of the glass at certain points during its cooling.

What Are the Properties of Ceramics?

The properties of materials determine their suitability for a particular job. Properties also govern

- how long the material will last in a particular use
- how well the material will hold up under adverse conditions
- how well the material will withstand harsh chemicals
- how well the material will withstand high temperature.

How Different Are Ceramics from Metals?

Ceramics are very different from metals in their properties. The mechanical, electrical, magnetic, and chemical properties of ceramics differ from those metals. To give you an example of how they differ, let's compare the metal aluminum with the corresponding ceramic, aluminum oxide (Al_2O_3). Aluminum oxide is chemically stable in a wide variety of environments, but the metal aluminum (Al) is not; aluminum would oxidize to aluminum oxide at higher temperatures and corrode in the presence of acids or bases. Aluminum oxide has a melting point of 2020°C (3668°F) and can withstand very high temperatures, but aluminum cannot. Aluminum has a melting point of 660°C (1220°F). The properties of aluminum oxide make it a popular material for heat-resistant, or refractory, purposes. It is widely used in the construction of high-temperature furnaces and as containers for the synthesis of several other ceramic materials.

You might wonder after hearing about the advantages of aluminum oxide why we don't use aluminum oxide instead of aluminum in automotive engines. The answer to this question lies in the differences in mechanical properties between aluminum oxide and aluminum. The ductile and malleable properties of metals allow us to stretch, squeeze, or bend them with tensile, compressive, or flexural forces. Ductility permits the metal to withstand fairly severe impact loading without breaking. By contrast, ceramics such as aluminum oxide are brittle. They often break under pressure instead of bending. The presence of positively and negatively charged ions in ceramics does not allow them to bend or deform the way metals do.

Another reason ceramics tend to break is related to their **porosity** (having pores or openings). Pores and other microscopic flaws are created during ceramic processing. When forces are concentrated upon the area of the flaw or opening, breaking often results. This tendency to brittle breaking is the major limitation to the use of ceramics in certain applications. New ceramic processing techniques are being developed to increase the strength of ceramics and overcome these limitations.

Ceramics are electrically different from metals too. Metals are good conductors of electricity, but ceramics are generally poor conductors (and therefore good insulators) of electricity. For example, using Table 3-2, you can compare the electrical resistivity—measured in ohms—of the metal aluminum and the ceramic aluminum oxide. (Resistivity is the opposition to the flow of electrons in one centimeter of material.) Aluminum has a resistivity of 10^{-6} ohms per cm at room temperature, whereas the resistivity of aluminum oxide is 10^{16} ohms per cm. The difference in electrical behavior arises mainly from the differences in chemical bonding in the two materials. Aluminum has metallic bonding. Its electrons in the outer orbitals are not bound to any individual aluminum atom and are referred to as "free" electrons. These free electrons move easily under the influence of an electric field and result in good electrical conductivity. By contrast, the ceramic, aluminum oxide, has ionic bonding. The electrons in the outer orbital of the aluminum atom have been transferred to oxygen to give aluminum ions and oxide ions. These electrons are tightly bound to oxygen and aluminum and are unable to move easily under the influence of an electric field.

Table 3-2. Comparison of the Properties of Two Metals and Their Oxides

Property	Rhenium Re (metal)	Rhenium oxide ReO_3 (ceramic)	Aluminum Al (metal)	Aluminum oxide Al_2O_3 (ceramic)
Reactivity	Reacts with oxidizing acids	Oxidized by HNO_3 to $HReO_4$	Easily oxidized to Al_2O_3. Reacts with acids and bases	Chemically stable
Melting point	3180°C (5756°F)	Decomposes at 400°C (752°F)	660°C (1220°F)	2020°C (3668°F)
Mechanical	Ductile	Brittle	Ductile Malleable	Brittle
Electrical resistivity	10^{-4} ohm/cm Conductor	10^{-6} ohm/cm Conductor	10^{-6} ohm/cm at room temperature Conductor	10^{18} ohm/cm at room temperature Insulator

Table 3-2 summarizes some of the properties of the metals aluminum and rhenium and the ceramics aluminum oxide and rhenium (rē′nē-əm) oxide. The table shows that the ceramic Al_2O_3 is more stable chemically, melts at a higher temperature, is an electrical insulator, and is brittle. The metal aluminum is not chemically stable, melts at a low temperature, is an electrical conductor, and is malleable and ductile.

Activity 3-4

- You are a manufacturer who needs a material to line the inside of a furnace. The material will be exposed to a temperature of 850°C (1562°F) and should be an electrical insulator. Study Table 3-2 and choose a material for this application.
- If the application called for an electrical conductor that could operate at a temperature of 250°C (482°F), which material would you choose?

Do not assume because aluminum oxide is an insulator that all ceramic materials are insulators. Some ceramics have an electrical conductivity similar to that of metals.

Other ceramic materials are semiconductors. Each ceramic material must be considered individually to determine if it is an insulator, semiconductor, conductor, or superconductor.

The Special Properties of Superconductors

Some materials lose all their electrical resistance below a certain temperature. These materials are called superconductors. The temperature below which they change to superconductors is called the critical temperature, or T_C. Above the T_C, the material has electrical conductivity similar to conductors. Figure 3-9 shows the relationship between resistivity and temperature.

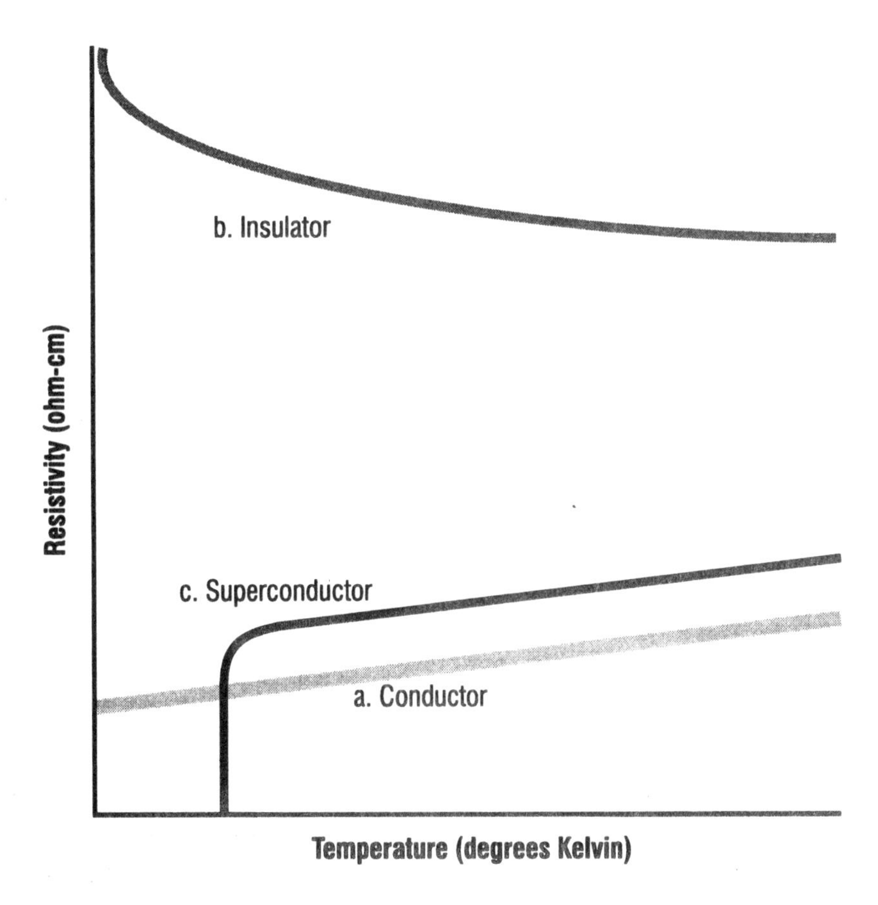

Figure 3-9 Change of resistivity with temperature for conductors, insulators, and superconductors

Activity 3-5

CAUTION: LIQUID NITROGEN CAN BURN THE SKIN. DO NOT LET THE LIQUID NITROGEN CONTACT YOUR SKIN. WEAR GOGGLES AND GLOVES WHILE POURING.

- Get from your teacher three circular disks—one of the superconductor $YBa_2Cu_3O_7$, one of steel, and one of stone. Place each disk in a petri dish. Place a tiny permanent magnet on top of each disk. Observe what happens with each disk and record your observations in your *ABC* notebook.
- Remove the magnets. Pour liquid nitrogen into each petri dish until the disk is completely covered and allow five minutes for the disks to cool. Add more liquid nitrogen as needed.
- Use the tweezers to place the permanent magnets onto the disks again and observe what happens. Record in your *ABC* notebook any differences between what happened with and without the liquid nitrogen.
- Discuss as a class which materials behaved differently at liquid nitrogen temperature and at room temperature, and which materials behaved the same at both temperatures.

As you have observed, superconductors exhibit a peculiar behavior in the presence of a magnetic field. Figure 3-10 compares the behavior of a superconductor and a nonsuperconductor in the presence of a small magnet. The magnetic field can penetrate a nonsuperconducting metal and the magnet just rests on the nonsuperconductor due to the **gravitational force**. By contrast, the magnetic field does not penetrate a superconductor. Instead, the changing magnetic field sets up a current in the superconductor. This current meets no electrical resistance and therefore continues. The magnetic field from the current opposes the magnetic field of the magnet. This causes the magnet to seem to float above the superconductor! This was first observed in 1933 by a scientist named Meissner and is called the Meissner effect.

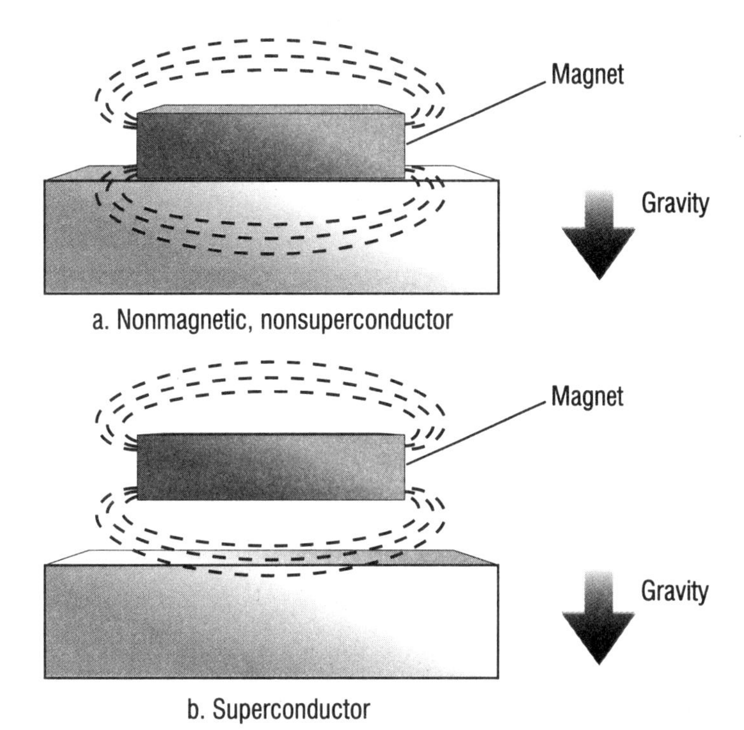

Figure 3-10 Magnetic interactions with nonmagnetic nonsuper-conductors and superconductors

How is a superconducting current different from a nonsuperconducting current? Three physicists—John Bardeen, Leon Cooper, and Robert Schrieffer—provided an answer to this question in 1957. Their answer is called the BCS theory and states that, at low enough temperatures, the electrons pair up and move in a cooperative manner without colliding with atoms or other electrons.

The normal conductor is similar to an unorganized crowd where each individual is moving independently. Even if the individuals in the crowd are moving in the same direction, there is a lot of bumping and jostling, causing the overall progress to be slow. The superconductor is similar to an army marching in step. Even though a large group is moving, the movement is quick and efficient because all the individuals are in step. This idea is shown in Figure 3-11.

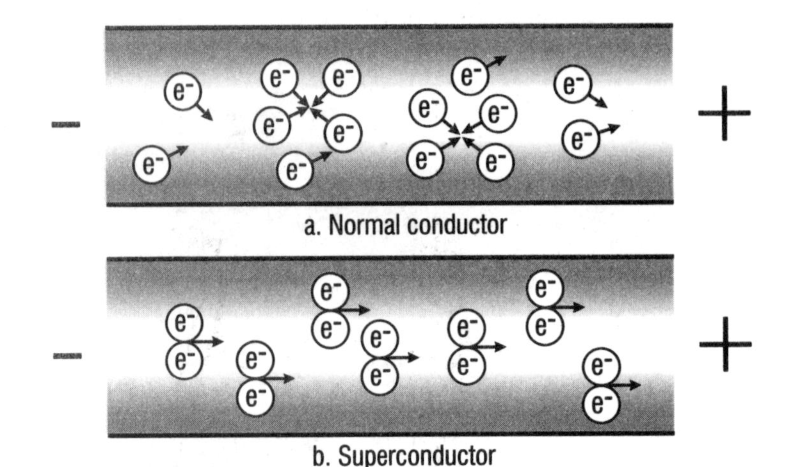

Figure 3-11 Electron movement in a nonsuperconducting current and a superconducting current

The BCS theory helps to explain superconductivity at very cold temperatures. The recent development of superconductors at warmer temperatures is hard to explain by the BCS theory. Physicists are still searching for a better explanation for the "warm" superconductivity behavior.

What Is Important About Ceramic Superconductors?

Before 1986, the highest known critical temperature was about 20K and the only known superconducting materials were metals and their alloys. In 1987, superconductivity was observed at a record high temperature of 93K in the ceramic $YBa_2Cu_3O_7$. This temperature is about 180 C° (324 F°) below the freezing point of water, but about 16 C° (29 F°) above the boiling point of liquid nitrogen as shown in Figure 3-12. The discovery of superconductivity at warmer temperatures allows the use of cheaper liquid nitrogen as a coolant instead of the more expensive liquid helium. This lower cost makes applications of superconductivity practical.

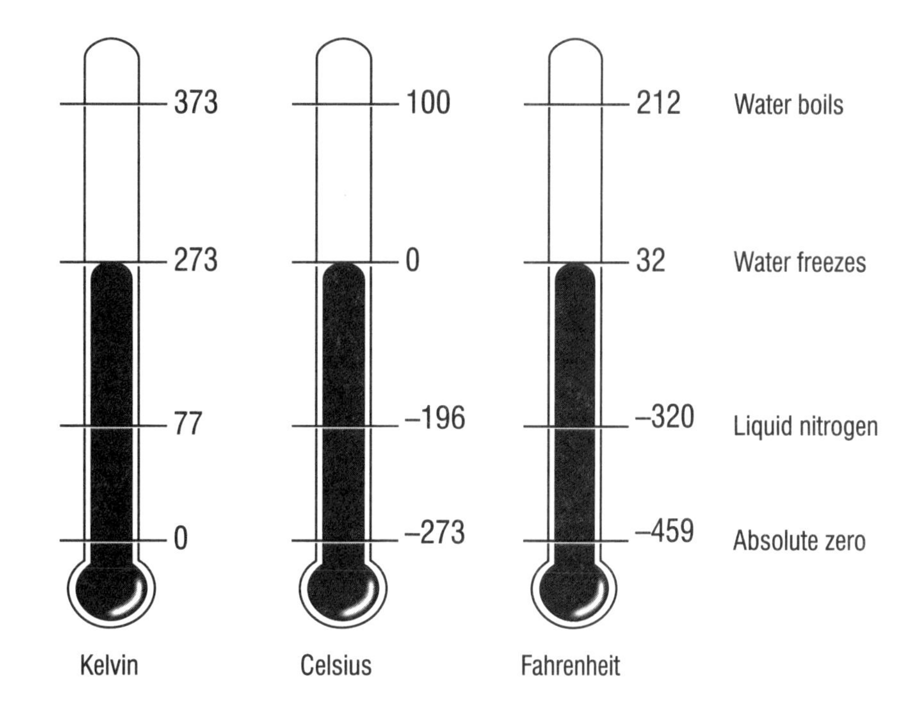

Figure 3-12 Comparison of temperature scales

The discovery of liquid-nitrogen-temperature superconductivity stimulated research in laboratories all over the world. As a result of this intensive research effort, a critical temperature, T_c, to only about 125K is possible in the ceramic $Tl_2Ba_2Ca_2Cu_3O_{10}$. Current research is focused both to raise T_c and to find revolutionary applications for these ceramic superconductors. Some anticipated applications of ceramic superconductors include

- Electric power storage and transmission without energy loss from resistivity
- Superconducting magnets, generators, or electric motors to generate power
- Magnetically levitated trains that can have speeds up to 300 miles per hour
- Ultrafast computers that can do trillions of calculations per second
- **Magnetic resonance** (rĕz′ə-nəns) **imaging (MRI)** to view the human body by noninvasive means.

However, the brittle property of ceramic superconductors poses several problems in these possible applications.

Magnetism in Ceramics

An electron spins around its own axis in addition to orbiting around the nucleus. This spinning can be compared with a current moving in a very small loop; it can generate a magnetic field. The orbital motion of the electron around the nucleus also generates a magnetic field. The magnetic field from the orbital motion is called the **orbital magnetic moment**. The magnetic field from the electron spin is called the **spin magnetic moment**.

The electron spin can be oriented so that the momentum is up or down—"spin up" or "spin down." In an electron configuration diagram such as Figure 3-13 for the nitrogen atom, the "spin up" is represented by an up arrow and the "spin down" is represented by a down arrow. In Figure 3-13, the 1s, 2s, and 2p represent different regions of space or orbitals where the electrons have different energy values.

The 1s orbital electrons have the lowest energy. The 2s orbital electrons are the second-lowest-energy electrons. The 2p orbital electrons are the highest-energy electrons in a nitrogen atom with the lowest-possible-energy electrons. Notice that there are three 2p orbitals. These represent three regions of space where the electrons that occupy them have equal energy. Electrons in the different types of orbitals, that is s or p for nitrogen (d and f orbitals are also possible) have different orbital magnetic moments.

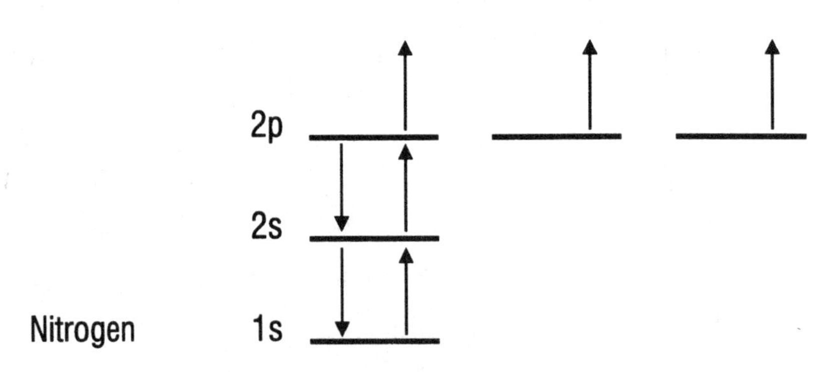

Figure 3-13 Electron configuration diagram for a nitrogen atom (Pauli exclusion principle)

When the atoms of a material such as neon have orbitals completely filled with electrons, all the electrons are paired and there is a cancellation of both the spin and the orbital moments and the material has no net magnetic moment. When these atoms are placed in a magnetic field, they are repelled by both poles of the field and align perpendicular to the lines of force. These materials are called diamagnetic.

When the atoms of a material have unpaired electrons, the material can have a net magnetic moment. These materials are attracted to a magnetic field and are called paramagnetic. In paramagnetic materials, the individual magnetic moments are randomly oriented at temperatures above 0K due to thermal agitation. However, in the presence of a magnetic field, the moments tend to align parallel to the applied magnetic field. In some materials, the magnetic moments on individual atoms may be aligned in a particular manner in the absence of a magnetic field even at temperatures substantially above 0K. There are three different possibilities of ordering as shown in Figure 3-14:

- Ferromagnetism—all the spins are aligned parallel to each other.
- Antiferromagnetism—spins are aligned in opposite directions in equal numbers.
- Ferrimagnetism—spins are aligned in opposite directions, but in unequal numbers.

At high temperatures, the ordering is lost due to thermal agitation and the materials become paramagnetic.

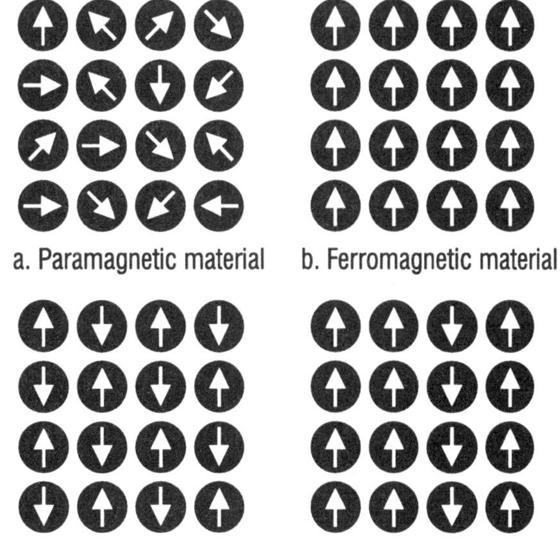

Figure 3-14 Ordering of individual magnetic moments in various types of materials

Antiferromagnetic ceramics do not have a net magnetic moment and do not have magnetic applications. Both ferromagnetic and ferrimagnetic ceramics have net magnetic moments and are technologically useful. For example, CrO_2 is a ferromagnetic ceramic and Fe_3O_4 is a ferrimagnetic ceramic. Both of these ceramics are used as coatings for magnetic recording media such as audio tape, video tape, and computer disks.

Dielectric Ceramics

When an electric field is applied to an insulating material, there may be a separation of positive and negative electrical charges to form what is called an electric dipole. These materials are called **dielectric** (dī′ĭ-lĕk′trĭk) materials and the formation of an electric dipole is called **polarization** (pō′lər-ĭ-zā′shən). In some ceramic insulators, permanent electric dipoles may exist on a molecular or atomic level as shown in Figure 3-15. These electric dipoles may be randomly oriented at temperatures above 0K due to thermal agitation, as in the case of paramagnetism in which magnetic moments also are randomly oriented. In the presence of an electric field, the dipoles will tend to align parallel to the applied magnetic field.

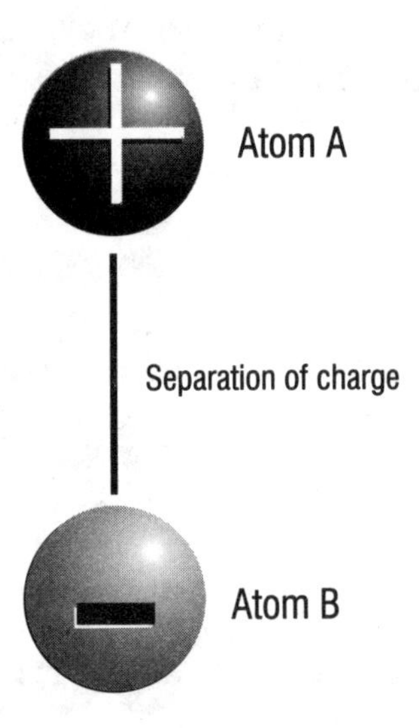

Figure 3-15 Representation of dipoles in a crystal

In a few ceramics, the electric dipoles may be aligned in the same direction even in the absence of an electric field—as in ferromagnetism—to give a ferroelectric ceramic. $BaTiO_3$ is a ferroelectric ceramic at temperatures below 120°C (248°F).

In a few other ceramics, polarization can occur when the ceramic is distorted by pressure. This is called piezoelectricity, which means pressure electricity. $PbZrO_3$ and $NH_4H_2PO_4$ are **piezoelectric** (pī-ē′zō-ĭ-lĕk′trĭk) ceramics.

What Kinds of Products Are Made from Advanced Ceramics?

CAREER PROFILE: RESEARCH SCIENTIST

Dr. M. is a professor at a major state university. He teaches courses in materials science and engineering. He does research in the area of ceramic materials. He trains students in the synthesis, processing, chemical characterization, and property measurements of ceramics.

One of the primary research interests of Dr. M. is to understand the relationship between crystal structures and the properties of ceramic materials. "Ceramics exhibit a wide range of useful properties that are controlled by the arrangement of atoms in the ceramic—this is the crystal structure—and the nature of interaction among the atoms," says Dr. M. He takes the simplest example, carbon, and elaborates upon the importance of this concept. "Carbon occurs primarily in two different forms: the expensive diamond, and the relatively cheaper graphite. These two forms have very different properties. Diamond is hard and a poor conductor of electricity whereas graphite is soft and a good conductor. These differences arise directly from the differences in the arrangement of carbon atoms in the two materials." He goes on, "Innovative laboratory synthesis has led to a recent discovery of a third form of carbon called fullerene that has a molecular shape similar to the shape of soccer balls."

He emphasizes that it is important to understand why certain ceramics are, for example, superconductors while most other ceramics are insulators. "One needs a strong foundation in the chemistry and physics of ceramics to find answers to such questions," says Dr. M. He further explains, "An understanding of the structure-property relationship is the key to designing new materials that can perform specific engineering functions.

"Equally important in controlling the properties and performance of ceramics are the synthesis and processing conditions one employs," Dr. M. notes. "The particle size plays an important role in ceramics properties. Ceramics with particle sizes of the order of a few nanometers

(1 nanometer = 10^{-9} meter) are referred to as nanophase ceramics and they exhibit extraordinary properties that are otherwise unachievable with conventional ceramic materials. **Sol-gel processing** is a novel technique to obtain nanophase ceramics." Dr. M. and his students are involved in the synthesis and processing of several technologically important nanophase ceramics and ceramic-metal composites.

Another area of Dr. M.'s interest is to identify ceramic materials that can be used as electrodes in rechargeable lithium batteries. He explains, "Lithium batteries are lightweight and are attractive for portable devices, but the reactivity of lithium metal with the electrolytes has hampered the development of commercially available rechargeable lithium batteries." He is optimistic that suitable lithium-insertion ceramics such as lithium-inserted carbon may provide a solution to this problem.

The wide variety of properties allows ceramic materials to be used in a variety of ways. Many new ceramic materials are being developed for new uses each year.

Advanced ceramics find a wide range of applications in modern technology with specific properties and functions. Some example functions and the corresponding properties and applications are shown in Table 3-3.

Advanced ceramics can be classified into two subcategories based on the type of application—functional or structural. Functional ceramics are those used to perform a specific function such as electronic, magnetic, optical, mechanical, nuclear, chemical, biological, opto-electronic, or electromechanical. Structural ceramics are those used to make structural parts such as engine components, valves, bearings, heat exchangers, cutting tools, or hip implants.

As Table 3-3 shows, there are many advanced ceramic applications, and to explore them all is beyond the scope of this subunit. Here we will look at two applications: oxygen sensors in automobiles and ceramic materials used as implants in the human body.

Table 3-3. Functions, Properties, and Applications of Advanced Ceramics*

Function	Properties	Example Ceramics	Applications
Thermal	Refractoriness, insulation, heat collection, thermal conductivity	SiC, Si_3N_4, ZrO_2, muillite ($Al_6Si_2O_{13}$), cordierite ($Mg_2Al_4Si_5O_{18}$)	High-temperature industrial furnace lining, electrode material, heat sink for electronic parts
Mechanical	High strength, wear resistance, low thermal expansion, lubrication	WC, SiC, Al_2O_3, ZrO_2, TiC, Si_3N_4, BN, SIALON, ZrO_2 + Al_2O_3, Al_2O_3 + ZrO_2, TiC + TiN, WC coated with TiC or BN, manufactured diamond	Tools and jigs, abrasives, turbine blades, solid lubricants, precision instrument parts
Biological	Biological compatibility	Calcium phosphate, $Zr(HPO_4)_2{\cdot}H_2O$	Artificial bone and tooth
Chemical	Catalysis	SiO_2, Al_2O_3	Fixed catalyst-carrier
	Electrodes and electrolytes	TiS_2, $LiCoO_2$, $Zr_{100-2x}Y_{2x}O_{200-x}$	Batteries and fuel cells
	Transport	$Zr_{100-2x}Y_{2x}O_{200-x}$, $MgCr_{2-x}Ti_xO_4$,	Sensors
Electrical	Electrical insulation, electrical conductivity, semiconductivity, piezoelectric, dielectric	Al_2O_3, AlN, BeO, SiC, $BaTiO_3$, $SrTiO_3$, $(Ni, Mn)_3O_4$, $KTaNbO_3$, $BaTiO_3$, $LiTaO_3$	Integrated circuit substrate, resistance heating element, varistor sensor, piezoelectric filter
Magnetic	Magnetic	Fe_3O_4, CrO_2, $Y_3Fe_5O_{12}$	Memory element
Optical	Optical condensing, fluorescence, translucence, optical conductivity	$A1_2O_3$, MgF_2, ZnSe, SiC, $LiNbO_3$, $LiTaO_3$, Ruby (Cr_3^+: $A1_2O_3$), silicates	Laser diode, light emitting diode, heat-resistant translucent porcelain, optical communication cable
Nuclear	Radiation resistance, refractoriness, high-temperature strength	SiC, Si_3N_4, ZrO_2, muillite ($Al_6Si_2O_{13}$), cordierite ($Mg_2Al_4Si_5O_{18}$)	Nuclear fuel, nuclear fuel cladding, control material, moderating material, reactor lining

*Adapted from an illustration developed by the Fine Ceramics Office, Ministry of Industrial Trade and Industry, Tokyo, Japan.

Ceramics Used for a Chemical Function—Oxygen Sensors

The oxygen sensor $Zr_{100-x}Y_xO_{200}$ is used in automotive engine control devices to reduce emissions and fuel consumption. Optimum emission control and fuel economy occur at an air-to-fuel ratio of approximately 15 to 1. The zirconia (zûr-kō′nē-ə)-based oxygen sensor monitors the ratio and provides a feedback signal to the control system to adjust the ratio to the optimum setpoint. In this application, use is made of the oxide-ion conduction property of $Zr_{100-2x}Y_{2x}O_{200-x}$. Substitution of a small amount of yttrium (ĭt′rē-əm) ions for zirconium ions in ZrO_2 creates oxygen vacancies to maintain a neutral charge. For example, in a typical oxygen sensor about 20% of the zirconium is substituted with yttrium, giving a formula of $Zr_{80}Y_{20}O_{190}$ and resulting in about 10% oxygen vacancy in the ceramic. These vacancies allow oxide ions to move through the ceramic to give oxide ion conductivity.

Substituting yttrium in the ceramic in order to change the properties of the ceramic (and therefore the performance characteristics of the oxygen sensor) is a good example of how ceramic materials are designed to have specific properties. Often materials engineers may change the percentage of an element or a compound in a ceramic. To do this, they must calculate the new chemical formula for the desired ceramic. You can practice this in the next activity.

Activity 3-6

- In oxide ceramics, zirconium, yttrium, and oxygen exist as Zr^{4+}, Y^{3+}, and O^{2-}. Assume that 50% of the zirconium in ZrO_2 is replaced by yttrium. What would be the chemical formula for the new ceramic?
- Begin with an empirical formula of $Zr_{100}O_{200}$. Replacing 50% of the zirconium ions would give 50 zirconium ions and 50 yttrium ions. The total positive charge for the 100 ions is $(50 \times 4 + 50 \times 3) = 350$. There should be one oxygen for every two positive charges to balance the charge in the formula; therefore, there must be $(350/2) = 175$ oxygen ions. The formula for the ceramic is $Zr_{50}Y_{50}O_{175}$. A general formula for this is $Zr_{100-2x}Y_{2x}O_{200-x}$.
- What would be the formula of a ceramic if 40% of the zirconium ions in ZrO_2 were replaced by yttrium ions? What would be the formula for each of the following percentage replacements of zirconium with yttrium—60%, 30%, 70%, 80%, 20%?

Ceramics Used for a Biological Function—Bone and Hip Joint Replacement

A natural ceramic substance called hydroxyapatite (hī-drok′sē-ăp′ĭ-tīt′) makes up about 75% of the weight of the bones of a vertebrate animal. Fibers of protein make up the other 25%. Hydroxyapatite reinforced with protein results in bones with several important properties. Bones are strong, resilient, and resistant to impact. Therefore, they can support the weight of our bodies and provide for the movement of our limbs (arms and legs). Some bones function mainly in protecting certain organs of our bodies from damage. What organs in your body are protected by bone?

Activity 3-7

Examine models of bone or actual animal bones that have been dissected to show an internal view. Use a stereomicroscope if available. What pattern of structure is common to all of the bones you have examined?

Hydroxyapatite, by itself, is a nonliving material—a matrix of various salts, especially calcium phosphate. But bone is alive. It contains living cells, a supply of blood, and a network of nerves. When you examined the cross section of bone in Activity 3-7, you probably saw at least two different regions of bone. Bone tissue with a spongy appearance is found near the ends of bones, where the bones take the most stress. Solid or compact bone is found throughout the length of the bone and surrounds the spongy areas at the end. Such areas of compact bone are rigid and not very flexible. Bones also contain areas of soft connective tissue called **cartilage** (kär′tl-ĭj). As you grow older, much of the cartilage in your bones accumulates calcium phosphate and hardens to become bony tissue (see Figure 3-16).

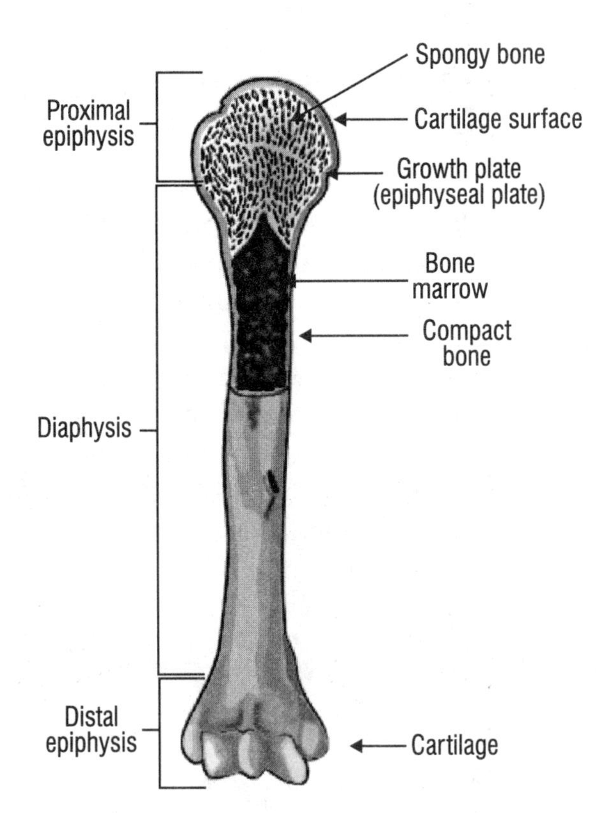

Figure 3-16 Cross section of bone

Synthetic Hydroxyapatite for Bone Replacement

Hydroxyapatite that is nearly identical to that found in bone can now be cheaply synthesized in the laboratory. Therefore, there is a demand for it as a biomaterial. A biomaterial is a nonliving substance that is used to take the physical place of or replace the function of living tissues. For applications such as reconstruction of the jaw, hydroxyapatite has proved to be very biocompatible. A biocompatible material is one that interacts with the body in a harmonious way.

Activity 3-8

- Hydroxyapatite has the **chemical formula** $Ca_5(PO_4)_3OH$.
- In the periodic table, find the **atomic weight** of the elements that make up hydroxyapatite.
- Then, using the chemical formula, calculate
  - the percent of calcium in bone
  - the percent of phosphorus in bone
  - the percent of oxygen in bone
  - the percent of hydrogen in bone
  - the ratio of calcium to phosphorus in bone.

An important consideration for an orthopedic surgeon who performs bone implants is that bone is a growing tissue: Bone grows as cells located within the cartilage of bone tissue take up calcium and phosphorus from the blood. When the cells deposit these minerals within the bone, the bone increases in length and width.

Bone cells may also remove minerals from the bone, a process called resorption. This can occur when a person fails to exercise. Movement of the muscles attached to a bone puts stress on bone, which helps prevent resorption.

Biomaterials to be implanted into the skeleton must take the normal processes of bone growth and resorption into account. Materials that have a rough surface area allow the body's own bone tissues to surround and attach to the implanted material. This, in turn, allows some of the stress applied to the implant to be transferred to the bone, which helps keep the bone healthy. Ceramics such as hydroxyapatite and a new type of glass ceramic called bioglass are currently considered the most biocompatible materials for skeletal implants.

Biomaterials for Joint Replacement

A joint is the junction between two or more bones that allows for movement of body parts. A typical joint, the knee, is shown in Figure 3-17. As you can see, the knee is a joint between the thigh bone (femur), the calf bone (tibia), and the kneecap (patella).

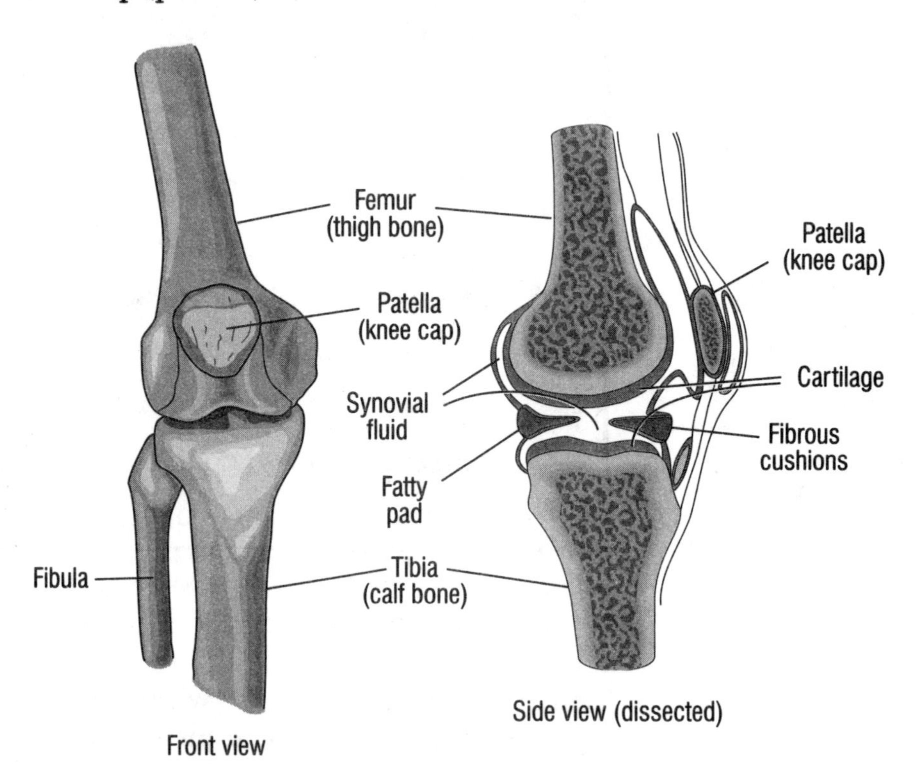

Figure 3-17
Knee joint

But a joint is much more than just the junction point of bones. The bones of a joint are connected by fibrous tissue including connective tissue called ligaments. The ends of the bones are covered with a wear-resistant layer of fibrous tissue called cartilage. Between the layers of cartilage of adjoining bones is a small cavity coated with a thin film of fluid, which has the consistency of egg white. This fluid, called synovial fluid, lubricates the cartilage layers of the joint. Various fibrous and fatty areas around the joint help to cushion the joint and distribute the body's weight onto it.

Movement at a joint occurs when there is contraction of a muscle that is attached to bones on either side of the joint. For instance, contraction of a muscle in the thigh causes the knee joint to flex so that the lower leg moves backward toward the back of the thigh (Figure 3-18).

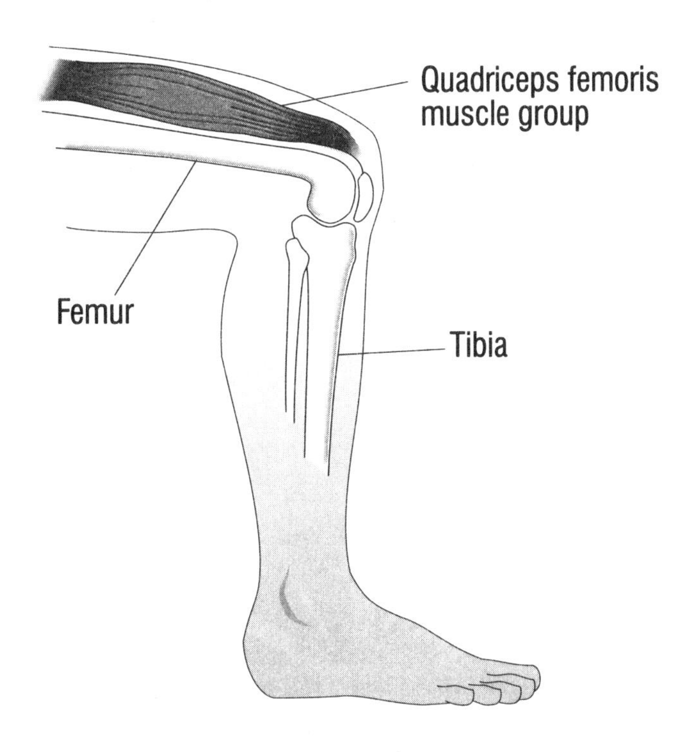

Figure 3-18 Flexion of knee joint

The four major joints of the human skeleton are shaded in the diagram of Figure 3-19, the shoulder, elbow, hip, and knee. The shoulder and hip joints are alike in that they are typical ball-and-socket-type joints. The ball-and-socket joint allows for a wide range of movement, especially rotational movement. The elbow and knee joints are somewhat alike in that they are both complex joints involving hinge-like and gliding movements. The elbow, in addition, provides some degree of rotational movement.

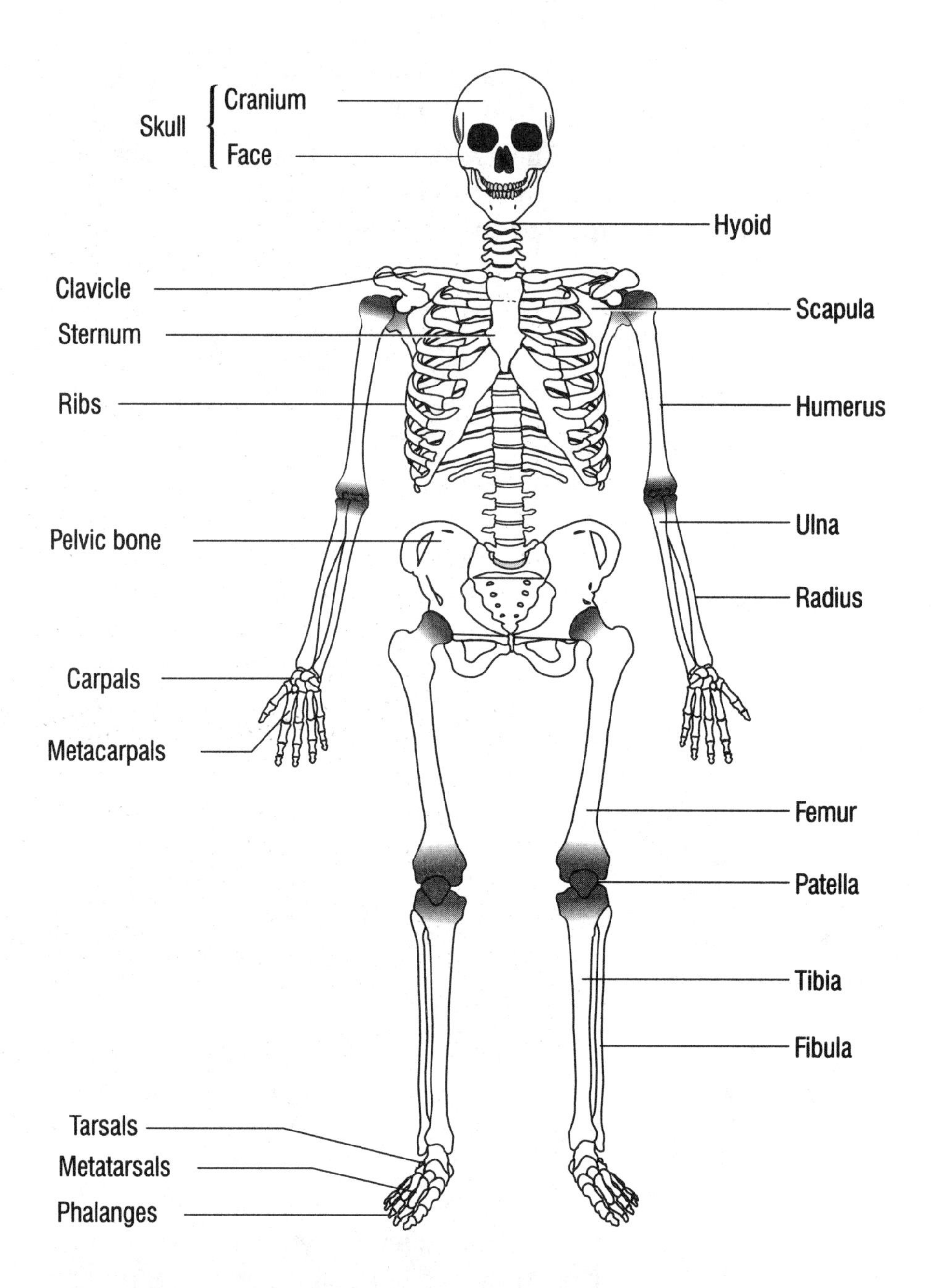

Figure 3-19 Human skeleton

Joints in the body are subject to various disorders, such as **arthritis** (är-thrī′tĭs), and to damage from injury or overuse. When a joint is no longer functioning adequately, an orthopedic surgeon may recommend the removal and replacement of the joint. A common joint replacement is the hip-joint replacement, in which the ball of the femur and the socket of the hip bone are surgically removed and replaced with synthetic parts. In the past, metal parts have been frequently used as the ball and mounting rod in an artificial hip joint. Ceramics such as alumina and zirconia are more often assuming this role today because of their greater resistance to wear and corrosion and their better biocompatibility.

The material used for a joint implant must not inhibit natural processes of the bone on which it is mounted. If bone recedes from the boundary of the implant, the implant loosens and does not function properly. If this happens, a second implant surgery becomes necessary.

Activity 3-9

- In the preceding sections you looked at two specific areas of application for ceramics. Return to Table 3-3 and select one of the applications to investigate. Assume that you are interested in a sales job for a company that makes ceramic for your selected application. You have an interview scheduled in two days, and you know very little about the company's product.
- Using technical dictionaries and/or encyclopedias, find out what properties are most critical for this application as well as any information about how the product is made. Write a brief report in your *ABC* notebook, explaining what you found out about the product. You will use this report in the Unit Wrap-Up Activity.

Manufacture

Traditional ceramic products are manufactured by processing naturally occurring raw materials such as clays and rocks. In contrast, advanced ceramics require controlled composition and structure and are usually synthesized in laboratories using artificial raw materials.

How Are Ceramics Made?

Laboratory synthesis of ceramics often requires chemical reactions between two solids. For example, the ceramic SiC is made during the reaction of silica (SiO_2) and coke (C) at a high temperature (1500°C) sustained for several hours in a nonoxidizing atmosphere such as argon.

$$SiO_2\,(s) + 3C\,(s) \longrightarrow SiC\,(s) + 2CO\,(g)$$

(silica) (coke [carbon]) (silicon carbide) (carbon monoxide)

Equation 3-2

The complex superconducting ceramic $YBa_2Cu_3O_7$ is made at 950°C (1742°F) in air by the following reaction—

$$2Y_2O_3 + 8BaCO_3 + 12CuO + O_2 \longrightarrow 4YBa_2Cu_3O_7 + 8CO_2$$

(yttrium oxide) (barium carbonate) (copper oxide) (oxygen) (superconducting ceramic) (carbon monoxide)

Equation 3-3

What is noticeable about these two example reactions? Both require high temperatures. One requires a special nonoxidizing atmosphere, whereas the other can take place in air. You may wonder why high temperatures and (in the case of silicon carbide) atmospheric controls are necessary.

Why Do Some Ceramics Need Very High Temperatures and Longer Reaction Times?

The answer to this question relates to the movement of atoms and molecules. **Diffusion** (dĭ-fyo͞o′zhən), the net movement of atoms or molecules in response to concentration differences, is usually slow in solids. Ceramic raw materials usually are made of particles that are about 10^{-5} meters in diameter as shown in Figure 3-20. Each particle consists of about 100,000 atoms or molecules. Chemical reactions between particles of raw materials thus requires diffusion to take place over a relatively long distance. However, in a solid, even at high temperatures, diffusion over the necessary distance is difficult to achieve. Another way to overcome the problem of slow diffusion is to allow long reaction times.

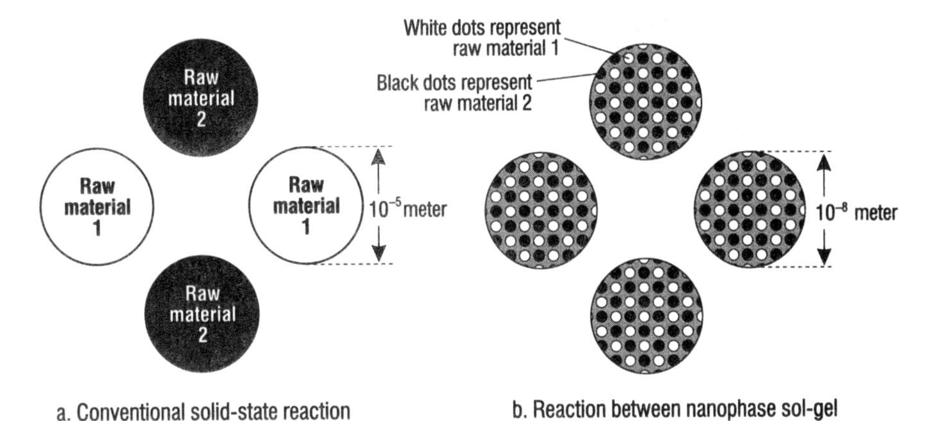

Figure 3-20 Comparison of particle size in solid phase reactions

Why Do Some Ceramics Require Atmospheric Control?

The answer to this question about atmosphere relates to the desired oxidation state of the product. If the desired product is a nonoxide ceramic such as SiC, an air atmosphere may not work. Oxygen in the air would oxidize the silicon and carbon atoms. Therefore, an inert atmosphere—argon gas—must be used. If the desired product is highly oxidized, an air atmosphere is suitable. In the case of the superconductor $YBa_2Cu_3O_7$, the air supplies the oxygen needed in the reaction.

The following generalizations are helpful:

- Synthesis of oxide ceramics, where the metal is to be highly oxidized, is carried out in air as in ferric oxide, Fe_3O_4. However, if a lower oxidation state is required as in ferrous oxide, (FeO) a nonoxidizing atmosphere must be used.
- Nonoxide ceramics such as carbides, silicides, nitrides, phosphides, sulfides, and selenides generally need a nonoxidizing atmosphere.

New Approaches in Ceramic Synthesis

The repeated grinding and heating process used in ceramic manufacturing is tedious, and new methods are being developed. One example is the manufacture of the oxygen sensor, $Zr_{100-2x}Y_{2x}O_{200-x}$. Conventional synthesis would require mixing ZrO_2 and Y_2O_3 powders and heating at higher temperatures (>1000°C or 1832°F) with repeated grinding. But, it can be made at much lower temperature (<800°C or 1472°F) with shorter firing time by a solution-based technique called the sol-gel process. This process is represented in Figure 3-21.

Here's how the sol-gel process works. When ammonium hydroxide is added to a solution of Zr4+ and Y3+, solid particles of zirconium hydroxide and yttrium hydroxide precipitate simultaneously from the solution. The precipitate is a homogeneous mixture of zirconium hydroxide and yttrium hydroxide. This process leads to a mixing of ZrO_2 and Y_2O_3 particles at a much finer scale. (The scale difference is as shown in Figure 3-20 with Figure 3-20 representing sol-gel synthesis powders.) The individual particles are about 10^{-8} meters (10 nanometers) in diameter and are called nanophase materials (1 nanometer is 10^{-9} meters).

The distances the reacting molecules must travel before they can react (called diffusion length) in nanophase materials are reduced about 1000 times. This reduction in diffusion length results in a shorter reaction time and a lower temperature. The sol-gel process also helps to fabricate the ceramic and improve its strength and other properties.

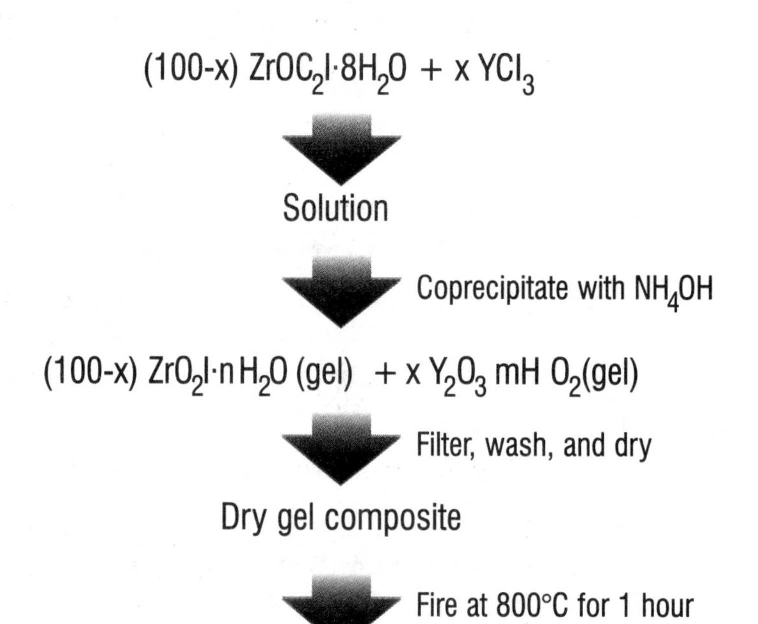

Figure 3-21
Sol-gel process for $Zr_{100-2x}Y_{2x}O_{200-x}$

Several other new techniques for ceramic synthesis have been developed, including vapor-phase synthesis using lasers and the **pyrolysis** (pī-rŏl′ĭ-sĭs) (changing chemically through heat) of polymer precursors. For example, the pyrolysis of polysilane polymer (Figure 3-22) yields the ceramic SiC.

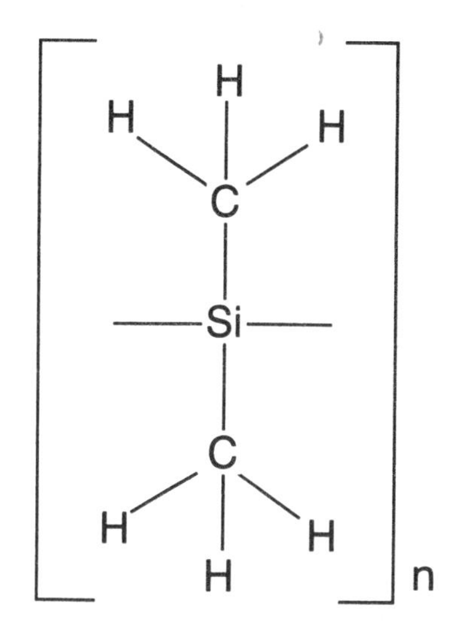

Figure 3-22 Monomer unit or polysilane polymer

How Are Ceramic Shapes Made?

CAREER PROFILE: CERAMICS TECHNICIAN

Ida M. is a ceramics technician for a company that makes electronics instruments. Ida produces a ceramic material in powder form. Working according to the specifications of a supervising materials engineer, Ida keeps detailed records, including microphotograph, of all her procedures in the laboratory.

Ida's primary task involves mixing the basic elements used to make the ceramic with water, certain acids, and other chemicals. After mixing, she submits this mixture to what is called a "dirty burn," a heating process that removes impurities. She crushes the purified mixture into a fine powder. Mixing and purifying this batch of power is only the first step, however. Ida then must test the powder to determine if it has unwanted materials in it, and improve the next mixture based on the test results. She then tests the new mixture to verify its electrical properties, and continues until she has produced a batch that exhibits the required properties.

Ida's job requires that she use specialized equipment such as autoclaves (a device for sterilizing), a microscope with an attached camera, a specialized oven, grinders, and polishers, to name but a few. Ida has been working in the lab for over ten years and has become skilled in operating and maintaining it. Ida has a high school diploma and came into the company with good reading and math skills. She gained most of her job skills and knowledge from her 26 years of experience in the company.

However, she recommends that anyone who wants to be a ceramics technician should have at least a year of **inorganic** chemistry in college, preferably more.

You have been introduced in Subunit 2 to shape-forming processes that are used for metals and alloys. Metal shapes are generally formed by pouring the molten metal into molds. The same casting procedure does not work for ceramics because most ceramics decompose at or near the melting temperature. Ceramic shapes are formed by several other techniques referred to generally as green forming. The specific green-forming technique used depends on the shape to be formed, the nature of the ceramic being fabricated, the property to be achieved, and the production cost. Figure 3-23 shows the sequence involved in making most ceramic shapes.

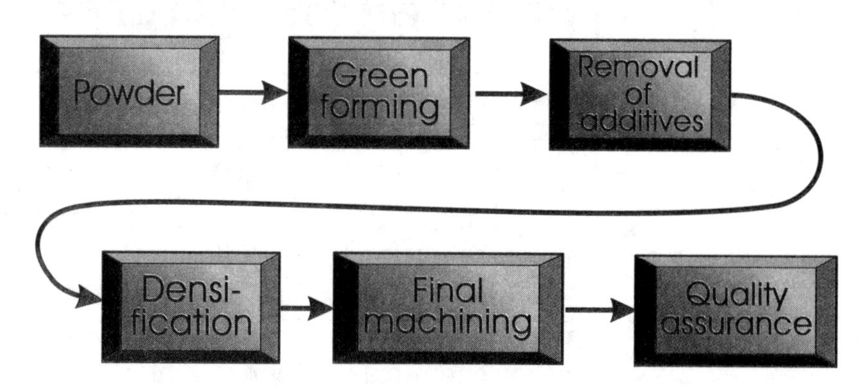

Figure 3-23 Shaping a ceramic material

Processing of Ceramic Powders

Many ceramic shapes are made from chemically synthesized powders or powders derived from naturally occurring raw materials.

The particle size is an important parameter in ceramic powder processing, and the desired particle size is generally achieved by crushing and milling procedures. The starting powders require some prior treatment so that they will flow. The type of treatment depends on the type of process and the material. Organic plasticizers, binders, lubricants, **deflocculents** (dē′flŏk′yə-lənts), wetting agents, and **sintering** (sĭn′tər ĭng) (a type of heat treating) aids are often added to the powder at this stage. Each additive has special functions.

Green Forming

Green forming is a term used to refer to the techniques used to shape a ceramic material. As listed earlier, these include powder pressing, slip casting, tape casting, **extrusion,** and injection molding. The result of the green-forming process is a shaped ceramic piece called the green body. The green body then undergoes further processing.

Activity 3-10

- Work in research teams of two to four people each. Let each team select one of the green-forming techniques listed in Figure 3-22 to investigate. Your teacher will provide you with a general description of green-forming techniques so that you will have some background and a list of ceramics companies. Your research may take the form of telephoning or writing for information, checking the library for background materials, or even touring local companies, if possible.
- Each team should make a presentation to the class that describes the green-forming technique. You may use diagrams, company brochures, or sample demonstrations to enhance your oral presentation.

Removal of Additives

The organic materials added to the powder during the powder processing step are removed from the green body by a suitable firing procedure. During this process, care should be taken in handling the green body to avoid any damage.

Densification

The green body formed is unusually porous (has openings or pores) and may not have adequate strength. The removal of porosity is called **densification**. The required strength can be achieved by one of several methods, including sintering, hot pressing, or hot isostatic pressing (HIP).

Sintering involves heating the porous green body at higher temperatures. The enhanced diffusion at higher temperatures eliminates porosity and leads to densification. The reduction in the surface area of the green body during sintering causes a reduction in the surface free energy. Green shapes that are small in size have a higher surface area and, therefore, undergo sintering at much lower temperatures and for shorter times. Shapes formed with fine powders usually give higher densification and strength. Chemical synthesis of powders plays an important role in this aspect. Sintering can occur by either solid-state diffusion or liquid-state diffusion. The former is called solid-state sintering and the latter liquid-phase sintering.

Liquid-phase sintering is faster than solid-state sintering. In most advanced ceramics, solid-state sintering occurs. In some cases, an intermediate liquid phase may be formed by reaction between the starting powders, and the liquid phase may assist the sintering process

In hot pressing, pressure is applied to the green body in a die at higher temperatures. Combination of high pressure and temperature enhances diffusion and thereby the sintering process.

Hot pressing is either mono- or bidirectional. In hot isostatic pressing, pressure is applied uniformly in all directions at elevated temperatures.

Final Machining

After the densification process, the ceramic part goes through a final machining: cutting and surface finishing. Final machining ensures that the precise dimensions and suitable surface finish are achieved. The part at this stage will be strong and have enough strength to withstand the machining forces. The high hardness of ceramics necessitates the use of hard diamond tools for machining, so this process sometimes becomes expensive.

Quality Assurance

The ceramic part is finally subjected to a series of tests to evaluate its properties and performance. The type of test depends on the type of application for which the ceramic part is made. The parts are often subject to nondestructive testing (NDT) to detect whether any surface defects or cracks are present. The presence of even microscopic defects can reduce the strength of the part. X-ray radiography is a common technique used for quality assurance in ceramics.

How Are Glass Ceramics Made and Fabricated?

Most commercial glass products contain SiO_2 and other oxides such as CaO, Na_2O, Al_2O_3, and B_2O_3. Glass ceramics are obtained by heating the required raw materials above the melting temperature and rapidly cooling the melt to room temperature. **Homogeneity** (hō′mə-jə-nē′ĭ-tē) in glass is generally achieved by complete melting and uniform mixing of the raw materials. SiO_2 readily forms glass for the following reasons. The polar bonds of SiO_2 are strong and directional and so the three-dimensional network of Si-O bonds does not readily break and reform upon cooling.

Glass ceramics are fabricated by pressing, blowing, drawing, and fiber-forming methods. Pressing is used to fabricate relatively thick-walled pieces such as plates and dishes. Drawing is used to form long pieces such as rod, tubing and fibers.

Annealing (ə-nēl′ĭng) (repeated heating and cooling) of most glass ceramics at higher temperatures can cause a transformation from noncrystalline to crystalline state. This process is called devitrification. This process is generally avoided because **devitrified** (dē-vĭt′rə-fīd′) glass is nontransparent. Also, the noncrystalline to crystalline transformation may be accompanied by volume changes, which may introduce stresses and weaken the material.

Manufacture of Semiconductors

One of the largest uses of ceramic materials is semiconductors for the electronics industry. These materials are manufactured under very carefully controlled conditions to give the desired products.

Looking Back

Ceramic materials are chemical compounds made of metallic and nonmetallic elements. The main types of bonding that occur in ceramic materials are ionic and covalent bonding. Ceramics may have a crystalline or an amorphous microstructure. Crystalline ceramics have a long-range repeating structure, but amorphous or glass ceramics do not.

Ceramics differ from metals in many of their properties. They are usually less reactive or corrosive, less malleable, and more resistive to electrical current, and they have higher melting points than metals. The ceramic materials known as superconductors can conduct current very quickly and efficiently, but most superconducting materials developed so far require unusually low temperatures to perform as superconductors.

Applications for advanced ceramics are being found in many fields. These applications take advantage of the improved thermal, mechanical, electrical and electronic, biological and chemical, optical, and nuclear properties of materials. For most advanced ceramics, the properties are tailored precisely to the application through careful control of the processes of making and forming the ceramic material.

Further Discussion

- What might be some of the problems associated with developing superconducting materials?
- Because ceramics tend to be brittle, they cannot be used for certain applications. What uses or applications do you think are likely not to be suitable for ceramic materials? What might be done to a ceramic material to

make it possible to use it in one or more of these applications?

- What may happen in 100 years to the numerous glass skyscrapers that have been built?

Activities by Occupational Area

General

Consumer Guide to Brick and Tile

- Contact a brick or tile company. Find out how the properties of tiles vary according to their different uses.
- Recommend certain tiles for the following locations in a home: entryway, kitchen, patio, and bathroom floor and walls.
- In your *ABC* notebook, make a simple diagram that provides a "consumer guide to tile" for others to use in making choices about tiles.

Agriscience

Windows on the Field

- Contact an agricultural equipment company and find out what kind of glass is used in the cabs of large tractors and other moving farm equipment. Does this glass have any properties that are different from that used in automobiles and trucks? Does it offer the user any protection from ultraviolet radiation?
- Assume you are an advertiser for the farm equipment company. In your *ABC* notebook, write a description of the features of the glass used that could be part of a sales brochure or ad.

Health Occupations

Ceramic Wave Guides

- Contact a biomedical equipment company or a hospital and find out if you can see a demonstration of the ceramic-wave guides used in microsurgery. If so, write a report about these in your *ABC* notebook.

Family and Consumer Science

Pottery Glazes

- Investigate pottery glazes used by one or more local potters. Find out if these are all suitable for objects used for eating and/or cooking. What kind of glazes are not suitable for eating and/or cooking utensils.
- Assume you are a consumer product specialist. In your *ABC* notebook, write a brief guide to consumers, explaining what questions to ask about glazes and what to look for to make sure that dishes and pots are safe to use.

Industrial Technology

Semiconductor Chip Manufacture

- Work in small groups to do this activity. Using research conducted in the library or from contacting semiconductor companies, develop a storyboard that shows the steps involved in making a semiconductor chip. Present your storyboard to the class.

LAB 5

HOW IS THE POROSITY OF BRICK MEASURED?

PREVIEW

Introduction

David C., the ceramics sales engineer introduced to you at the beginning of Subunit 3, supplies materials to many manufacturers of ceramics products. Many brick manufacturers are among his customers. David explains that, although bricks have been made for thousands of years, there is still a certain technical expertise required to deliver the right kind of brick to the customer.

"There are zones in the United States that are called 'freeze-thaw zones.' One of the fundamental evaluations of a brick or its suitability for use in a certain zone is the measure of its porosity," says David.

"If a brick is too porous, for example, it will absorb a high percentage of water. In a cold climate, the water inside the pores of the brick will freeze and later thaw. This freeze-thaw cycle eventually weakens the brick.

"To determine the capacity of a brick to absorb water, the lab weighs the sample brick, puts it in boiling water and leaves it there overnight, dries off the excess water, and weighs it again. The increase in weight indicates the amount of water absorption by the brick. This a simple test, but an important one, and it must be performed according to ASTM (American Society for Testing and Materials) standards."

Purpose

In this lab, you will measure the porosity of building brick.

Lab Objectives

When you've finished this lab, you will be able to—

- Compare the accuracy of two methods for determining the volume of an object.
- Determine the density of a liquid.
- Recommend how porous bricks should be for a given application.

Lab Skills

You will use these skills to complete this lab—

- Measure the weights of solids and liquids using a balance.
- Measure the volume of a liquid using a graduated cylinder.

Materials and Equipment Needed

unglazed building bricks, not > 4 cm thick, or unfinished clay tiles
piece of blackboard chalk
triple-beam balance
wire basket
metal tongs
mineral oil bath
bottle of mineral oil
bucket for mineral oil
10-ml graduated cylinder
500-ml plastic graduated cylinder
paper towels
newspapers
lint-free rag

Pre-Lab Discussion

One general property of ceramic materials is that they are porous (Figure L5-1). This property is important in the application of materials, such as bricks, which are sometimes exposed to a harsh environment. If a wet brick is subjected to freezing temperatures, ice forms in the pores. When the brick thaws, the ice expands as it melts. This freeze-thaw cycle eventually weakens ceramic materials structurally.

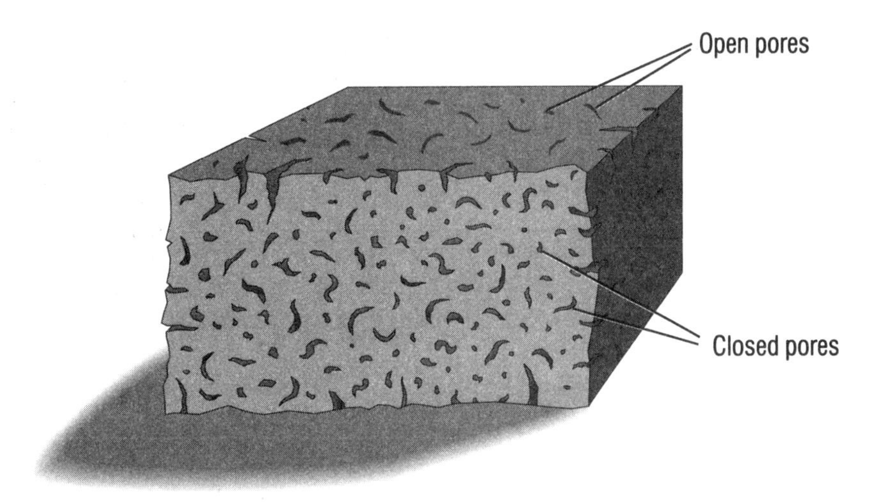

Figure L5-1
Pores in a typical brick

Substances other than water may penetrate a ceramic material and react chemically in such a way that the material becomes unsuitable for use. For some products, absorbance of unwanted substances can be prevented by coating the surface with a glaze. But the glazing of materials such as bricks would make them too expensive for use in construction.

Porosity is the measure of how porous a ceramic material is. Usually, a ceramic designer is interested in what percent of a material's total volume is made up by open pores (Figure L5-1). This characteristic is called apparent porosity. To determine apparent porosity, a ceramic technician needs to know both the volume of the brick and the volume of all its open pores.

For two reasons, apparent porosity cannot be measured directly. First, materials such as bricks have thousands of tiny open pores whose volumes would be impossible to measure by hand. Second, it is difficult to measure accurately the volume of an object, such as a rough-surfaced brick, from its external dimensions. Instead, this is done by submerging such objects in a liquid, and measuring their change in weight or the rise in the height of the liquid in a column.

In this lab, you will test the porosity of several kinds of bricks using a method similar to that prescribed by the ASTM for building brick.

Safety Precautions

- If you should spill mineral oil onto the floor, clean it up right away with paper towels so that someone does not slip and get injured.

LAB PROCEDURE

Method

Put on your lab apron and goggles.

Part A: Determining the Volume of Open Pores of a Brick

Day One

1. Obtain bricks of several different types. Using chalk, place a small distinguishing mark on each of them. Record this mark in column 1 of the data table.
2. Weigh each brick on a triple-beam balance. Record the weights in column 2 of the data table.

3. After you have weighed the bricks, use tongs to place each of them in the mineral oil bath. Be sure that the entire brick is covered by the oil.
4. Allow the bricks to soak in the oil overnight.
5. Determine the density of the mineral oil by doing the following:

   Weight a 10-ml graduated cylinder on the pan of the balance. Record this weight in your *ABC* notebook.

   - Remove the cylinder from the balance and fill it with 10 ml of mineral oil.
   - Wipe the outside of the cylinder dry and allow the oil to settle. Add (or pour off) oil if the level is less than (or greater than) 10 ml and wipe dry.
   - Reweigh the cylinder of oil. Record this weight in your *ABC* notebook.
   - Do Calculation A. Record the result in column 4 of the data table.

Day Two

6. Spread paper towels or newspaper on the lab counter. Using tongs, remove a brick from the mineral oil bath and place on the paper towels. Wipe all surfaces of the brick with a dry, lint-free rag.
7. Place a paper towel on the pan of the balance and tare the scale. Place the wiped brick in the pan and weigh. Record this weight as the saturated weight in column 3 of the data table. Do Calculation B.

Part B: Displacement Method of Determining Brick Volume

8. Fill a 500-ml plastic graduated cylinder with 250 ml of mineral oil. Holding the cylinder at an angle, allow one brick to slide gently into the oil and fall to the bottom of the cylinder. Read the new level of oil on the graduated cylinder. Record this level in your *ABC* notebook. Do Calculation C.
9. Using tongs, remove the brick from the cylinder and return to the mineral oil bath.
10. Repeat Steps A6-7 and B8-9 for each brick you weighed in Part A. When you have finished all pieces of brick, do Calculation D for each piece.

Part C: ASTM Method of Determining Brick Volume

11. Using tongs and a paper towel, transfer a brick from the mineral oil bath to the basket at the weighing station.

12. Lower the basket with brick into the bucket of mineral oil until the brick is completely submerged in oil. Be sure that the basket hangs freely from its supports and does not touch the bottom or sides of the bucket.
13. Balance the scale and record this weight in your *ABC* notebook. Your teacher will give you the tare weight (weight of basket and string). Do Calculation E. Record the submerged weight of the brick piece in Column 5 of the data table.
14. Using tongs and a paper towel, remove the brick from the basket and return it to the mineral oil bath.

15. Repeat steps C11-14 with each remaining brick you tested in Parts A and B. Do Calculations F and G.

Cleanup Instructions

- Wash graduated cylinders, tongs, and lab counters well with soap and water.
- Remove bricks from the oil, wipe dry, and place on paper towels.

Observations and Data Collection

Data Table

Number of brick	Dry weight of brick	Saturated weight of brick	Density of oil	Submerged weight of brick

Calculations

A. Find the density of the mineral oil from the following formula:

$$\text{Density of oil (g/ml)} = \frac{\text{Mass of oil and cylinder (g)} - \text{Mass of empty cylinder (g)}}{\text{Volume of oil in cylinder (ml)}}$$

B. Find the volume of open pores using the following formula:

$$\text{Volume of open pores (ml)} = \frac{\text{Saturated weight of brick (g)} - \text{Dry weight (g)}}{\text{Density of oil (g/ml)}}$$

C. Determine the volume of oil displaced by the brick using the following formula:

$$\text{Displaced volume (ml)} = \text{Final oil level (ml)} - \text{Initial oil level (ml)}$$

How does this displaced volume relate to the total (bulk) volume of the brick piece? Enter the bulk volume of the brick here: ________

D. Find the apparent porosity of the brick using the following formula:

$$\text{Percent apparent porosity} = \frac{\text{Open pore volume}}{\text{Bulk volume}} \times 100$$

E. Find the submerged weight of the brick using the following formula:

$$\text{Submerged weight of brick (g)} = \text{Total weight of brick and basket (g)} - \text{Weight of basket (g)}$$

F. Find the bulk volume of the brick using the following formula:

$$\text{Bulk volume} = \frac{\text{Saturated weight (g)} - \text{Submerged weight (g)}}{\text{Density of oil (g / ml)}}$$

G. Recalculate the apparent porosity of the brick using the bulk volume obtained in Calculation F.

WRAP-UP

Conclusions

1. For each part of the lab (A, B, and C), explain why it is necessary for the brick to be saturated with oil before measurements are made.
2. Explain the steps you took to calculate the density of the mineral oil in Part A.
3. Compare the percent porosity obtained in Part B (Calculation D) with that obtained in Part C (Calculation G). Which method of measurement is more precise? Explain.
4. Which of your brick samples would be most suitable for year-round use in your part of the country?

Challenge

5. Why was oil used to wet the brick in Part A instead of water?

LAB 6

OXYGEN CONTENT OF A CERAMIC SUPERCONDUCTOR

PREVIEW

Introduction

Dr. M., the ceramics research scientist who was introduced to you in Subunit 3, does not work alone in his laboratory like the mad scientists of the movies. Dr. M. is often found in his lab working side by side with his graduate students, teaching them techniques or offering help when it is needed.

Superconducting ceramics are the focus of much of the research in Dr. M.'s laboratory. One of the techniques carried out by Dr. M. and his graduate students is the determination of the oxygen content of a superconducting material.

"In several ceramic superconductors, the superconductivity of the material depends on its oxygen content. The superconducting temperature and the critical current density also depend strongly on the oxygen content of the material." Dr. M. explains that this technique is also carried out in the research laboratories of many electronics companies.

Dr. M. explains further that superconductivity is of interest to companies that are trying to develop superfast computers and information storage and retrieval systems. Some of Dr. M.'s students will likely be employed in industry research labs when their schooling is completed.

Purpose

In this lab, you will perform a redox titration to determine the oxygen content of a simulated superconducting ceramic solution.

Lab Objective

When you've finished this lab, you will be able to—

- Perform a redox titration to determine the oxidation state of an element with variable oxidation states.

Lab Skills

You will use these skills to complete this lab—

- Measure the volume of a liquid with a pipette.
- Measure the volume of a liquid with a graduated cylinder.
- Titrate a reaction mixture to a visually determined endpoint.

Materials and Equipment Needed

solution labeled "simulated $YBa_2Cu_3O_{7-\delta}$ solution 0.1 g/10 ml"
10% potassium iodide (KI) solution
0.05-N sodium thiosulfate ($Na_2S_2O_3$) solution
1% starch solution
3.5 N hydrochloric acid (HCl)
10-cc disposable plastic syringe (or 50-ml burette)
deionized water
250-ml beaker
3 250-ml Erlenmeyer flasks
10-ml pipette
pipette pump
50-ml graduated cylinder
10-ml graduated cylinder

Pre-Lab Discussion

The ceramic $YBa_2Cu_3O_{7-\delta}$, as people working with ceramics refer to these materials, is a superconductor below 93°K if $\delta = 0.06$, but an insulator if $\delta = 1$. The whole number ratio or empirical formula for this ceramic would be $Y_{50}Ba_{100}Cu_{150}O_{347}$ for the superconductor and $YBa_2Cu_3O_6$ for the insulator.

The ceramic material is synthesized by firing a mixture of 1 part Y_2O_3, 4 parts $BaCO_3$, and 6 parts CuO in air at 950°C. At 950°C, the value of δ is 1, giving the insulator

$YBa_2Cu_3O_6$. As the ceramic cools slowly from 950°C (1742°F) to room temperature, it can insert an oxygen atom up to a maximum value of $\delta = 0.94$. Therefore, allowing the ceramic to cool slowly produces the superconductor, while rapid cooling does not allow enough time for the oxygen to insert. Thus, rapid cooling produces the insulator. Production of the superconductor is critically dependent upon the production procedures.

It is important to be able to verify the oxygen content of the ceramic. This is accomplished by a redox titration. The metals in the ceramic are Y, Ba, and Cu. The metals Y and Ba have only one possible oxidation state (oxidation state is the charge of the ion)— +3 for Y and +2 for Ba. However, Cu can have oxidation states of +1, +2, and, under some conditions, +3. The variation in the oxidation states for Cu is the reason for the variation in the oxygen content of the ceramic. The redox titration uses I ions to reduce all the Cu ions to the +1 oxidation state, then oxidizes the I_2 molecules to determine how many electrons were involved in the reduction of the Cu ions. The reactions involved in this titration are shown in Equations L6-1, L6-2, and L6-3.

$$2Cu^{++} + 2I^{-} \rightarrow 2Cu^{+} + I_2$$

Equation L6-1

$$Cu^{+++} + 2I^{-} \rightarrow 2Cu^{+} + I_2$$

Equation L6-2

$$2S_2O_3^{--} + I_2 \rightarrow S_4O_6^{--} + 2I^{-}$$

Equation L6-3

A sample of the ceramic is weighed, ground into fine particles, and dissolved. Then, a source of iodide ions (I^-) is added to the mixture and the pH is lowered. This reduces all of the Cu ions to the +1 oxidation state and forms I_2 from the I ions that react. The thiosulfate ($S_2O_3^{--}$) then oxidizes the I_2 back to I^-. The indicator for the titration is a starch solution that forms a deep blue complex with I_2. The amount of $S_2O_3^{--}$ used in the titration is used to calculate how many electrons were used to reduce all the Cu ions to the +1 state.

In this lab, you will not work with the ceramic. Instead, you will work with a copper nitrate [$Cu(NO_3)_2$] solution that simulates a $YBa_2Cu_3O_{7-\delta}$ solution. You will perform the titration and the calculations to determine the value of δ for the sample.

Safety Precautions

- Be careful with the acid solution used in this lab. If you spill any of the acid, tell your teacher and get help in properly cleaning the spill. If you spill any of the acid on yourself or your clothing, immediately wash the spill under running water for several minutes.

LAB PROCEDURE

Method

Put on your lab apron and goggles.

1. Using the 50-ml graduated cylinder, get 35 ml of the "simulated $YBa_2Cu_3O_{7-\delta}$ solution 0.1 g per 10 ml."
2. Use the pipette to measure a 10-ml sample from the graduated cylinder. Be sure to use the pipette bulb to draw the sample into the pipette. Transfer the sample into a 250-ml Erlenmeyer flask. Measure two more samples into the two remaining Erlenmeyer flasks.
3. Empty the remaining solution into the waste jar. Rinse the graduated cylinder with deionized water and dry it.
4. Label the Erlenmeyer flasks "Sample 1," "Sample 2," and "Sample 3."
5. Pour 15 ml of the 10% KI solution, measured with the 50-ml graduated cylinder, into each of the three samples. Rinse the graduated cylinder with deionized water and dry it.
6. Pour 10 ml of 3.5-N HCl, measured with the 50-ml graduated cylinder, into each of the three samples. The samples will now be a yellow-brown color from the I_2 that is formed by the reaction.

7. Into a clean, dry 50-ml graduated cylinder, pour 50 ml of 0.05-N $Na_2S_2O_3$ solution. From this graduated cylinder, fill the 10-cc syringe. Be sure that no bubbles remain in the syringe.

- Hold the syringe with the opening up,
- Draw air into the syringe.
- Tap the side of the syringe to release all bubbles.
- Slowly push the air out of the syringe.
- Fill the syringe to the 10-cc mark.

8. Add approximately 1½ ml of the 1% starch solution to each samples. The samples are now a deep blue color from the iodine–starch complex.
9. Begin with “Sample 1” and slowly add the 0.05-N $Na_2S_2O_3$ solution from the syringe. As you add the $Na_2S_2O_3$ solution, the sample will lose the blue color.
   - Begin by adding about 1 ml of the $Na_2S_2O_3$ solution and swirl the mixture.
   - As the color of the solution after swirling becomes lighter, decrease the addition to ½ ml.
   - When the blue color becomes even lighter, add one drop at a time until the blue color is gone.
   - If the $Na_2S_2O_3$ solution in the syringe runs out, refill the syringe as in step 7.
   - Record the final volume of the $Na_2S_2O_3$ solution added to the sample in the Data Table. Be sure to add the 10 ml if the syringe had to be refilled.
   - Repeat this step for “Sample 2” and “Sample 3.”

Data Table

Sample Number	Volume of 0.05 N $Na_2S_2O_3$ in milliliters
1	
2	
3	

Cleanup Instructions

- Empty the copper solutions into the waste jar provided by your teacher.

- Wash and dry all glassware and return it to its proper storage location.

Calculations

A. Calculate the value for δ in the formula $YBa_2Cu_3O_{7-\delta}$ using the formula—

$$\delta = \frac{0.4 - 662.2 \times V_1 \times 0.05 \times 0.001}{0.2 - 16 \times V_1 \times 0.05 \times 0.001}$$

Where V_1 is the volume of sodium thiosulfate used in the titration.

B. Calculate the value for δ for all three samples.

WRAP-UP

Conclusions

1. With the positive charges equal to the negative charges in the $YBa_2Cu_3O_{7-\delta}$ before titration, calculate the average positive charge on the copper atoms if Y has a +3 charge, Ba has a +2 charge, and O has a –2 charge.
2. Calculate the positive charge on the copper atoms in $YBa_2Cu_3O_7$.
3. Calculate the positive charge on the copper atoms in $YBa_2Cu_3O_6$.

Challenge

4. Would this redox titration method work for the determination of the oxygen content in a ceramic with metal ions that had only one oxidation state (possible positive charge)?

SUBUNIT 4

POLYMERS

THINK ABOUT IT

- Which objects in the picture above could be made out of plastics or polymers?
- What are some of the advantages of plastic materials? What are the disadvantages?
- With so many different kinds of plastics available, what do these materials have in common?
- How do you think the plastic playhouse might have been formed? How might the Styrofoam cups have been formed?

SUBUNIT OBJECTIVES

After you complete this subunit, you will be able to—

1. Model the chemical bonding involved in a polymerization reaction.
2. Explain how the molecular structure of polymers affects their properties.
3. Distinguish between the structure of thermosetting and thermoplastic polymers and the properties and uses of each.
4. Demonstrate selected tests that are used to verify the physical and/or chemical properties of some polymers.
5. Compare and contrast the different ways in which polymers are formed.
6. Investigate the safety, health, and/or environmental problems associated with the manufacture of selected polymers.
7. Determine the approximate boiling point of a liquid.
8. Separate a mixture of liquids based on their boiling points.
9. Understand how nylon polymers are made.
10. Demonstrate a chemical process.
11. Understand the concept of cross linking
12. Understand the concept of hydrogen bonding.

PROCESS SKILLS

You will use these skills in lab—

- Measure a liquid volume with a graduated cylinder.
- Read a thermometer to determine an approximate boiling point of a liquid.
- Accurately measure the volumes of solutions with a graduated cylinder.
- Read a Material Safety Data Sheet (MSDS).
- Accurately measure solutions using a graduated cylinder.

How Are Polymers Used?

Although you may not realize it, you use polymers every day. Polymers include many materials we normally think of as plastics—drink cups, milk cartons, and plastic combs. Polymers also are used to make linoleum floors, automobile taillight covers, and even carpet. Why are polymers so widely used? The answer is often because they are very strong and durable, yet very lightweight. Also, they can be designed to have other special qualities, making them insulators, oxygen barriers, or wrinkle-resistant cloth, to name a few.

The uses of polymers are so varied that exploring all of them would be beyond the scope of this text. Therefore, in this subunit, we will focus on a specific use of polymers in the medical field—their use as implants in the body's **cardiovascular system**.

Polymers have been used as implants in the human body for many years. One of the first polymers to have this function was polymethylmethacrylate, better known as PMMA. PMMA was used in World War II as windshield material in fighter planes. Cockpit windshields, when damaged in combat, would sometimes spray tiny fragments of PMMA into the eyes of pilots. Military surgeons, while trying to remove these fragments, frequently reported that the PMMA fragments had not caused the eye tissue to become inflamed.

In other words, PMMA seemed to be biocompatible. (As you may remember from Subunits 2 and 3, a biocompatible material is one that interacts with the body in a harmonious way.) PMMA's biocompatibility eventually led to its use as an artificial lens in the eyes of patients whose natural lenses had developed cataracts (clouding of the lens or surrounding membranes). The lenses made of PMMA were able to carry out some of the functions of the natural lens and help these patients see again.

The biocompatibility of synthetic polymers has led to their use inside the human body to serve various functions. So far, the most important of these polymers have been developed for use in the cardiovascular system—the system of heart and blood vessels that delivers blood to tissues throughout the body. Tissues such as heart valves and blood vessels, when defective, can now be replaced by materials that are made largely of synthetic polymers like Dacron, Teflon, and polyurethane. These materials, when implanted into the body in certain forms, are biocompatible not only with the muscles and other connective tissues of the cardiovascular system, but also with the blood.

The Cardiovascular System: Heart and Heart Valves

Your heart is a fist-sized organ that lies within your chest cavity a few inches behind your sternum. (The sternum, or breastbone, is a flat bone whose lower end can be felt as sort of a knuckle in the middle of your chest.) The heart is contained inside a special sac coated with a lubricating fluid that helps reduce friction in the sac when the heart pumps. Each day, the heart beats more than 100,000 times as it pumps the equivalent of some 2,000 gallons of blood to body tissues.

The human heart, like the hearts of other mammals, is made up of four chambers and four valves (Figure 4-l). The chambers are lined with strong muscles that, when they contract, force blood out of the chambers. These same muscles relax to let blood fill the chambers. On each side of the heart, a valve controls the passageway from the upper to the lower chamber. On each side of the heart, another valve controls the passageway between the lower chamber and the large blood vessel leading away from the lower chamber.

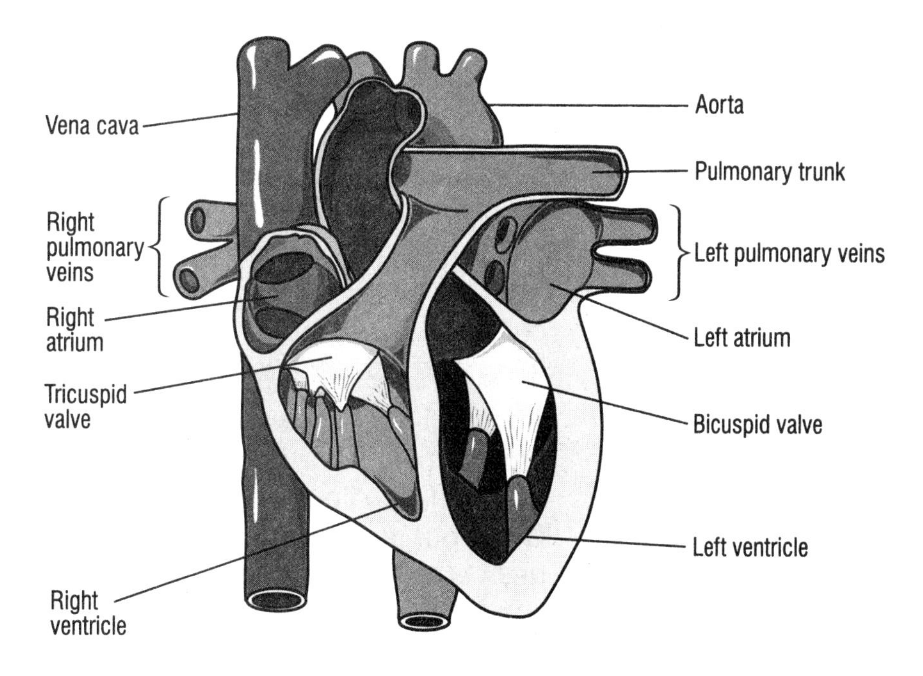

Figure 4-1
Human heart

The upper and lower chambers on the same side of the heart work in concert pumping blood. The chambers on the right side of the heart work together with the right-side heart valves to pump blood to the lungs, which lie close on either side of the heart. The chambers on the left side of the heart work together with the left-side heart valves to pump blood received from the lungs to tissues throughout the body.

Activity 4-1

- Work together in small groups. Examine a model of the human heart. On the model locate the following heart chambers:
  - Right upper chamber (right atrium)
  - Right lower chamber (right ventricle)
  - Left upper chamber (left atrium)
  - Left lower chamber (left ventricle)
- Now, locate the following heart valves:
  - Valve between the right atrium and ventricle
  - Valve between the right ventricle and large vessels (pulmonary arteries) leading from the right ventricle
  - Valve between the left atrium and ventricle
  - Valve between the left ventricle and large vessel (aorta) leading from the left ventricle
- Using a pointer or a marking crayon, indicate on the heart model the direction in which blood flows through the right side of the heart and through the left side of the heart.
- Working with other students in your group, determine how the function of the heart in pumping blood to the lungs and to other body tissues depends on the coordinated activity of each chamber and valve. To do this, your group must decide at what point during the heartbeat the muscles in the atrium walls and in the ventricle walls must contract and expand. Also determine when each of the four heart valves must open and close.
- After you have made your decisions, prepare an oral presentation using your model of the heart to explain the action of heart chambers and valves in pumping blood. Give your presentation to the class. Be prepared to answer questions your teacher and other students in the class may have.

Heart Valve Replacement

The valves of the heart sometimes may be defective, or may be damaged as a result of infection. When this happens, valves may not close or open fully. If a valve does not close adequately, there will be a backflow of blood, usually from the ventricle to the atrium. If a valve does not open fully, too little blood will be pumped to the ventricle, or too little will be pumped from the ventricle to the blood vessels leaving the heart. Because of reduced blood flow, patients with severe heart valve damage cannot participate in even moderate activity. They cannot climb stairs, carry bags of groceries, or get emotionally overexcited.

Through advances in biomaterials, however, some patients with heart valve disease may undergo surgery to remove and replace a defective valve. Various types of artificial heart valves are used as replacements (Figure 4-2).

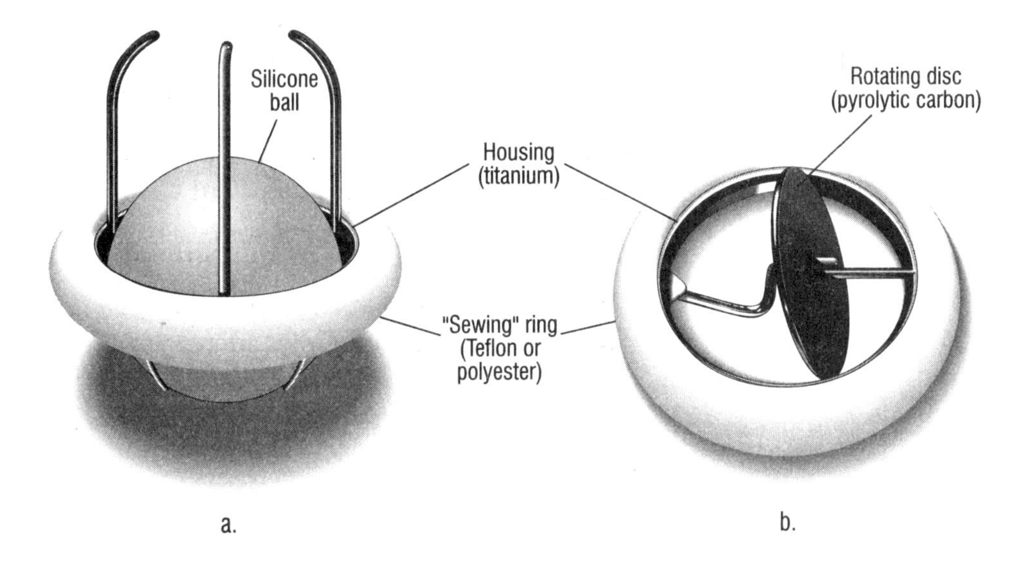

Figure 4-2 Artificial heart valves

Artificial heart valves or other implants needed in the cardiovascular system have several requirements:

- They must be very durable. If they begin to wear out, the patient will need another operation, which always involves risk. If the valves should stop functioning altogether, the patient's life will be threatened.
- They must not interfere with the normal flow of blood.
- They must not adsorb (attract and hold on their surfaces) suspended substances the blood such as proteins, lipids, and blood cells.

- They must not cause adverse chemical reactions in the heart, blood vessels, or the blood itself. Most important, an implant must not cause the blood to clot.

Clotting of Blood

Blood **clotting** is a complex chemical reaction that occurs in the blood when cell particles called **platelets** circulating in the blood respond to an irregular surface, such as a damaged blood vessel or the surface of a foreign object. When clotting occurs, a fibrous substance is deposited on the inside walls of blood vessels. The fibrous network traps blood cells in its vicinity to form a clot. The clot serves as a temporary repair of a damaged blood vessel.

With blood clotting there is always a danger that the clot will break loose and flow with the blood. The larger a moving blood clot, the more likely that it will completely block a blood vessel, a condition called an **embolism**. If an embolism occurs in the heart, the lungs, or the brain, the victim may die unless he/she receives emergency medical care.

A type of polyurethane used in the Jarvik-7 artificial heart (Figure 4-3) provides a smooth surface that is compatible with blood. Other synthetic polymers are biocompatible because of certain chemical reactions they trigger in the body. Artificial blood vessels made of Dacron or Teflon are designed to have irregular surfaces in contact with blood. An irregular surface causes cells in blood vessel walls to produce fibers of a protein called **collagen** (kŏl′ə-jən), which adhere to the implant. An implant "conditioned" by the body's natural response will be stable and will not be recognized as foreign material. It will not, therefore, trigger abnormal blood clotting or inflammatory responses such as swelling, reddening due to increased blood flow, and local temperature increase.

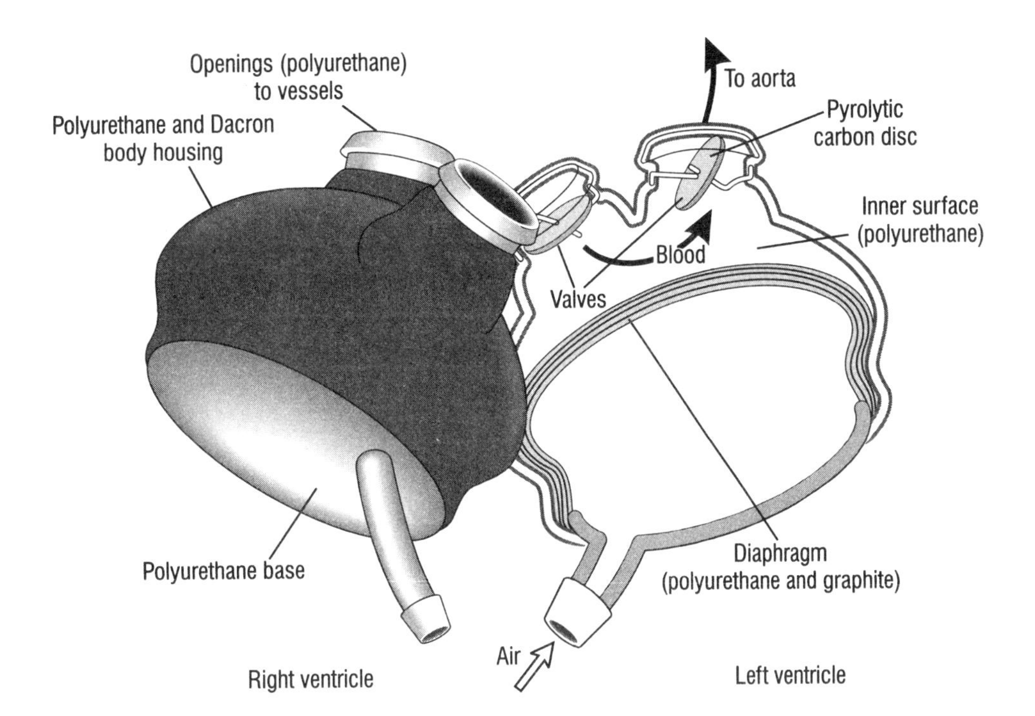

Figure 4-3 Jarvik-7 artificial heart

Types of Artificial Heart Valves

The first widely used artificial heart valves were of the caged ball variety (Figure 4-2a). The cage was made of stainless steel or titanium and had a tendency to make blood flow unevenly. The ball, the moving part of the device, was made of silicone. Silicone has a tendency to adsorb substances in the blood; therefore, its use in the **cardiovascular** (kär′dē-ō-văs′kyə-lər) **system** poses the threat of clotting. To reduce this danger, a medication called heparin is often prescribed. Although many patients with the caged-ball heart valves are leading active lives today, there are drawbacks to prolonged use of an anticlotting agent such as heparin.

A more recent and widely accepted artificial heart valve includes a rotating disc made of a polymer called pyrolytic carbon (Figure 4-2b). Not only does this type of valve encourage smoother flow of blood, but pyrolytic carbon is more durable and biocompatible than silicone. The part of the device that attaches to the patient's heart lining is a ring made of a knitted Teflon or polyester, both of which are biocompatible.

How Is a Polymer Made?

Polymers—from artificial heart valves to milk jugs—are made from **petrochemicals** (pĕt′rō-kĕm′ĭ-kəlz). Petrochemicals are chemical substances produced from oil and gas. If you have ever driven by a petrochemical plant, you know that they are often large and complex. A seeming maze of pipe extends high over plant buildings and tanks. Flames are seen shooting from tall flare stacks. From a distance, such plants seem like space-age cities.

Inside the gates of the plant, many different processes are being carried out. Crude oil is processed to make substances such as the chemicals called alkenes (**hydrocarbons** with a double covalent bond). Alkenes, in turn, are processed to make polymers. In the following section, you will have the opportunity to take an armchair tour of one part of a polymer plant.

Process Control—Tracking the Problem

Carmen is a process control operator at a polymer production plant. This particular polymer plant makes polyethylene, which is used to make many consumer ***products****.*

Carmen monitors the process equipment (see process flow diagram in Figure 4-4) to ensure that everything is running properly. Carmen notices that, for the last hour, the temperature in the raw material separator has been higher than normal. Now she notices the temperature rising in the product dryer as well. Carmen calls Hien, an instrument mechanic, and asks her to check the temperature gauges for accuracy, because she can't find a cause for the temperature increases. Hien reports that the gauges are working properly.

After another half hour, Carmen receives the analysis on the last sample of finished product polymer from Reggie, the lab technician. Reggie's results show that the finished product has a high level of impurities. Carmen thinks for a moment and decides that the raw materials being fed to the reactor must contain high levels of impurities, causing problems in the finished product. She knows that impurities can cause increases in pressures and temperatures in the process equipment.

Carmen calls the raw material supplier to tell them of a suspected problem with their material. Then, Carmen asks Jamie, the patrol operator responsible for the plant operation outside the control room, to start taking regular samples of the raw material until the problem is cleared up.

What Are the Steps in Polymer Production?

What is involved in the production of a polymer? For Carmen, the production process can be tracked by looking at the gauges and computer printouts in the control room. But what is behind the control room instrumentation?

The production process consists of several steps. Those steps and the equipment involved in a typical polymer production plant are covered in this section. Four of the steps deal with the polymer product and one with the raw material. The steps are

1. Chemical reaction—takes place in the **reactor**.
2. Separation of unreacted raw material from polymer product—takes place in the raw material separator.
3. Drying of polymer product—takes place in the dryer.
4. Finishing or reshaping of polymer into a form easy for customers to use—takes place in the finishing unit.
5. Raw material recovery or removal of impurities from unreacted raw materials—takes place in one or more distillation columns.

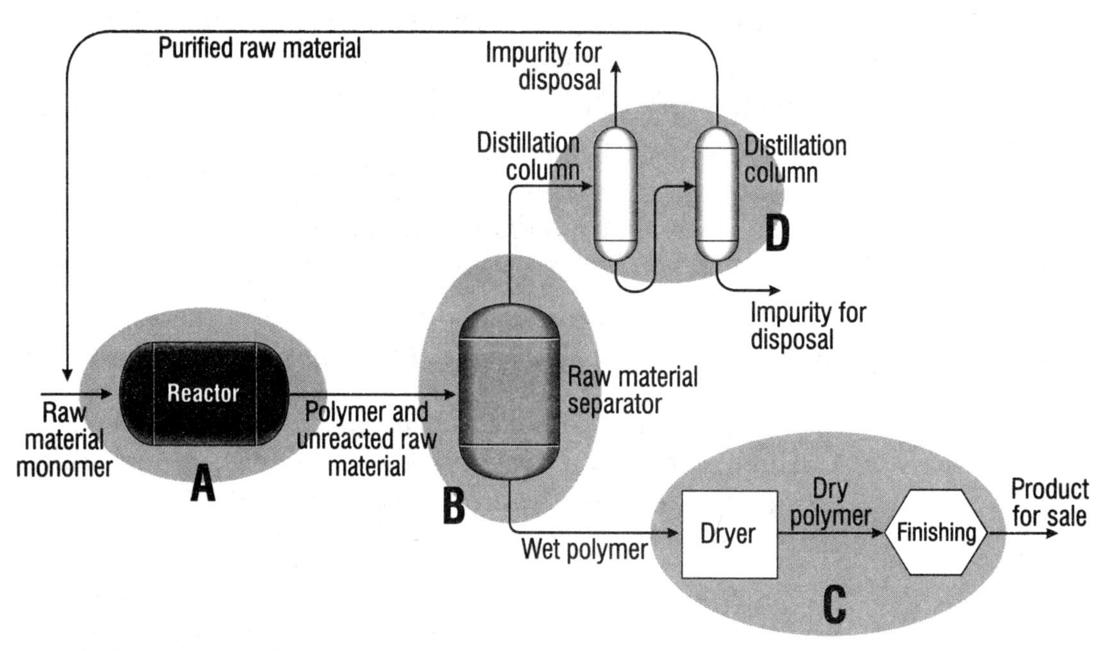

Figure 4-4
Polymer plant flowchart

Reaction, Separation, Drying, and Finishing

Step one in the production process is the reaction step. The reaction occurs in a piece of equipment called the reactor. The raw materials (reactants) are fed into the reactor (Figure 4-4a), where they form the product—the polymer. In a **polymerization** (pə-lĭm′ər-ĭ-zā′shən) reaction, the raw materials are called monomers.

Many of the polymers you use daily are solids. However, polymerization reactions usually take place at very high pressures and temperatures. These conditions cause the polymer to leave the reactor as a liquid or a gas. Both liquids and gases are called fluids. Fluid materials move more easily through the rest of the process equipment because they flow. It would be difficult to move a solid through the process equipment.

Some of the raw material fed into the reactor does not react. These unreacted raw materials leave the reactor with the polymer product. All the material leaving the reactor goes to the raw material separator (Figure 4-4b). The raw material separator removes the unreacted raw material from the polymer product. The raw material separator is heated and has a much lower pressure than the reactor. These conditions cause the raw material to **evaporate** and leave behind a solid polymer product. The raw material is then sent to the raw material recovery unit, and the polymer product proceeds to the product drier.

In the drying process, the now solid polymer moves through the dryer on a conveyer belt (Figure 4-4c). The dryer is hot enough to evaporate any liquid trapped in the polymer.

After the polymer has been made, separated, and dried, it is sent to the finishing process. In this step, the polymer product is melted and reshaped into a form that is easier for customers to use, usually pellets. Additives may be added to give the polymer special properties. This step is discussed in more detail later in this subunit.

Raw Material Recovery

The unreacted raw material that was separated from the polymer by the raw material separator goes to the raw material recovery unit (Figure 4-4d). This step doesn't involve the polymer; it processes the unreacted raw materials from the reactor to clean them. The cleaning process is done by removing any impurities from the raw materials in **distillation** (dĭs′tə-lā′shən) columns. The process often requires two distillation columns. Let's take a minute to learn the basics of distillation before we learn more about this part of the process.

A distillation column is a tall vessel designed to separate a mixture of materials. The column is fed with liquid flow, which enters around the middle of the column. The column is heated at the bottom, usually with steam. Heat is added until the temperature in the column is high enough to begin boiling the mixture. The material that boils **vaporizes** and begins to rise. The distillation column usually has many "trays" inside for the liquid to sit on. These trays have small holes that allow liquid to drop down and gas to flow up the column. The vaporized material rises and leaves the top of the column. The material that remains a liquid leaves through the bottom of the column.

Distillation columns work to separate materials, because different materials have different vapor pressures at a given temperature. The vapor pressure at a given temperature of a material is determined in part by its molecular weight and its **polarity** (pō-lăr′ĭ-tē) (the separation of charges in the molecule). However, if the polarities of two materials in a solution were too different, they would be unable to form a solution in the first place. In general, higher molecular weight and/or higher polarity result in a lower vapor pressure at a given temperature.

In a polymer plant, the first distillation column in the raw material recovery step removes low-boiling impurities. The unreacted raw material is fed to the column (see Figure 4-5). The temperature of this column is lower than the boiling point of the raw materials. Therefore, any impurity with a boiling point lower than the raw material's (low boiling component) boils. As the low-boiling impurity boils, it evaporates and leaves the column through the upper outlet. The raw material (and the remaining impurities) remain liquid and fall to the bottom of the column, where they exit through the lower outlet.

The unreacted raw material and remaining impurities go to the second distillation column (Figure 4-5). This second column removes high-boiling impurities. The temperature of this column is the boiling point of the raw material. As the raw material boils, it evaporates and leaves the column through the upper outlet. Any impurities with a boiling point higher than the raw material (high boiling impurities) remain a liquid and fall to the bottom of the column. These impurities can be removed from the column through the lower outlet for disposal.

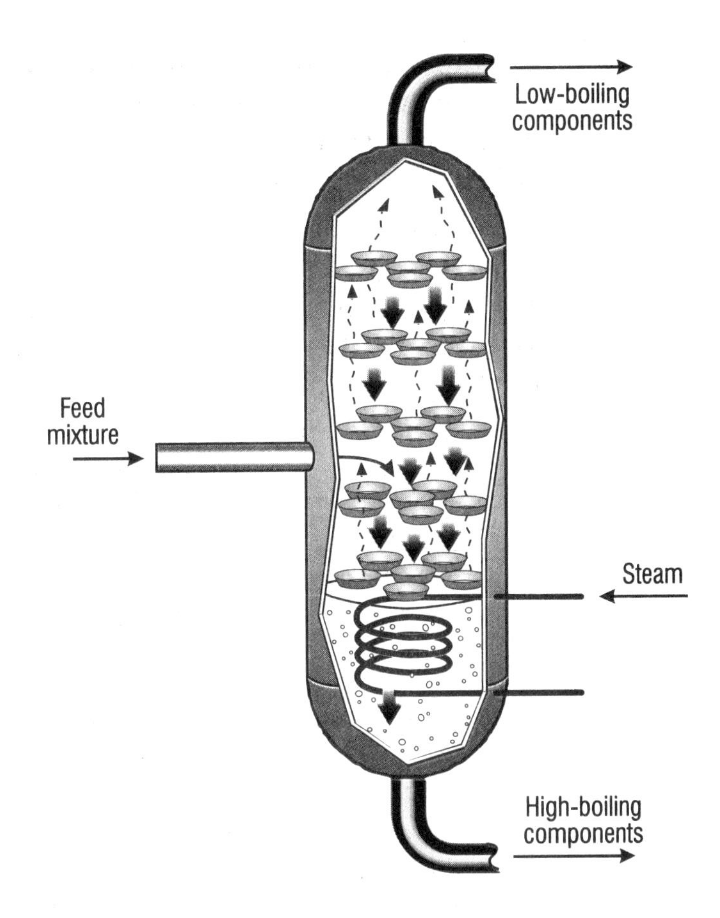

Figure 4-5
Distillation column

The raw material that left the top of the second column is now clean. It returns to the reactor for use in the reaction step again.

CAREER PROFILE: POLYMER PLANT OPERATOR

Carmen works as a process control operator at a polyethylene production plant. The plant purchases ethylene (a gas) and polymerizes it (explained in the next section) to produce polyethylene, a solid polymer.

Carmen spends most of her time in the control room. The control room has several computers that keep her informed about the process. She monitors flow rates, temperatures, and pressures inside process equipment, among many other things. These computers also allow Carmen to make adjustments to control the process. For example, if the process requires an increase in the temperature for a piece of equipment, Carmen tells the computer to increase the steam flow to the equipment, making it hotter.

Carmen earned an associate degree in chemistry. She must understand fundamental math and basic concepts of chemistry. She also needs to have practical knowledge of things such as electricity and mechanical equipment, since these are an important part of the production process. Her thorough understanding of the polymerization reaction and physical properties of the compounds with which she works is critical for the safe control of the process. Many other people in the plant depend on Carmen to safely run the process.

Activity 4-2

- Controlling a process, in which raw materials come into a plant and undergo physical and/or chemical changes to be formed into a product, goes on in many industries other than polymer companies. The processes may be very different in different plants, but some of the factors that are monitored are similar in many industries: pressure, temperature, and flow rate, to name a few. The instrumentation used to monitor these factors may also be similar.
  - Working with two or three other students, identify an industry in your community. It may be a small or large company. Arrange for a plant tour or a visit from a plant representative, or, if these options are not possible, ask to receive information about the process from the company.
  - Using the information you receive, develop a flowchart that shows the plant processes. Also make a list of the factors that are constantly monitored. Present your findings to the class.
- As a class, try to identify similarities and differences in the processes for these different companies.

What Is Polymerization?

What really happens inside the equipment we discussed earlier? Probably the most complex process occurs in the reactor. So far, we know that raw materials go in and the polymer comes out. Actually, it's a little more complicated than that! It is time to learn exactly how those raw materials react to become the polymer.

The raw materials are called monomers. The monomer enters the reactor and combines to form a polymer. (Mono means one; poly means many. Many monomers link together to form a long chain called a polymer.) Since the product of the reaction is a polymer, the reaction is called a polymerization reaction. There is more than one type of polymerization reaction. This section discusses two common types of polymerization reactions—the addition reaction and the condensation reaction.

Addition

The addition reaction is just what it sounds like—monomer molecules are added together end-to-end to build a very long chain of monomers—the polymer. What causes these monomers to link together? The reaction is started by adding a chemical known as an initiator to the raw materials (monomers) in the reactor. The initiator has an unpaired electron. The unpaired electron in the initiator makes it unstable. Molecules like the initiator are called **free radicals**. That single (unpaired) electron has a strong tendency to form a bond with another electron so it will be more stable. So the free radical takes an electron from a bond in a monomer molecule. It has to break one of the monomer's bonds in order to "steal" the electron from the monomer.

In the polyethylene polymerization reaction, the radical steals an electron from the carbon-carbon double bond in an ethylene molecule (see Figure 4-6). The double bond in ethylene becomes the target of the radical because that bond is more reactive than the carbon-hydrogen single bonds in the molecule. When the radical steals the electron from the double bond in ethylene, it forms a bond with one of the electrons in ethylene, and leaves a single electron "hanging open" on the ethylene where it broke the bond. Now the open electron on ethylene can form a bond! It takes an electron from a bond in another ethylene, attaching the ethylene to the end of the chain. Once again, it breaks a bond in the other ethylene and leaves an open electron at the end of the chain, which looks for another electron so it can form another bond.

This process continues many times, linking thousands of ethylene molecules to form the polyethylene polymer.

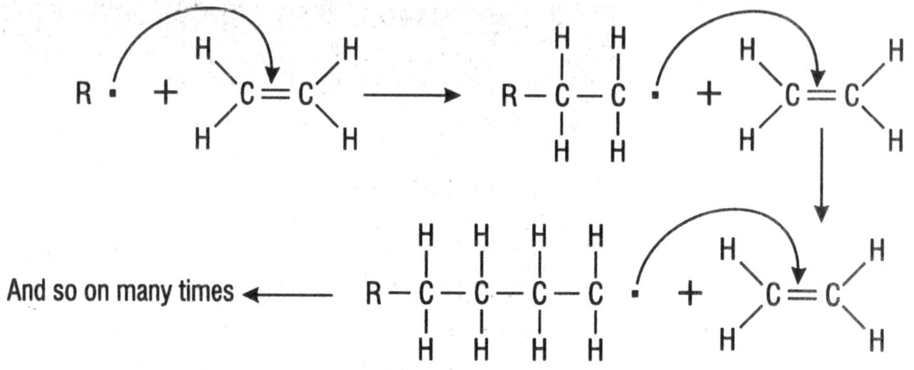

And so on many times

Figure 4-6
Polyethylene polymerization reaction

— Represents a bond formed by two electrons
· Represents the single electron in a free radical
R Represents a hydrocarbon chain made of carbon and hydrogen atoms

You might ask yourself: What keeps a polymer from becoming the blob that ate Chicago? In other words, what finally causes the chain to stop forming?

The linking process (polymerization reaction) stops when the unpaired electron on the end of the polymer chain comes in contact with another unpaired electron. The second unpaired electron can come from another polymer chain or from an initiator molecule. When two unpaired electrons come together, neither electron has to break a monomer bond to become stable. Instead, it bonds with the other unpaired electron. When these two unpaired electrons bond, no bonds are broken, so no electron remains unpaired. This step of the reaction is called the termination step, since no additional reaction can occur. Figure 4-7 shows the termination step using two polymer free radicals.

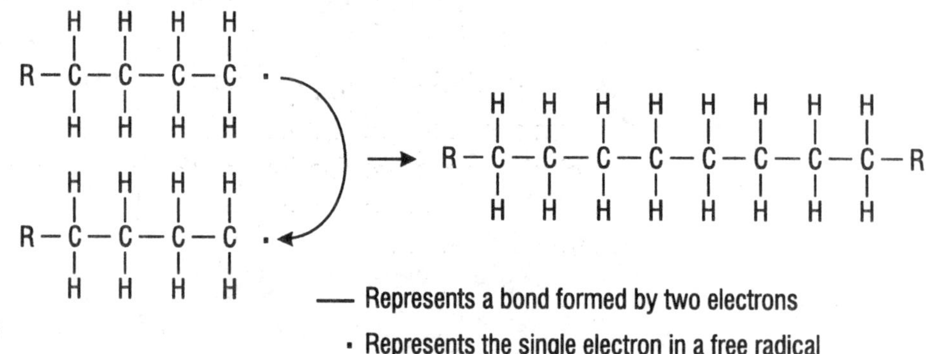

Figure 4-7
Termination of polyethylene polymerization reaction

— Represents a bond formed by two electrons
· Represents the single electron in a free radical
R Represents a hydrocarbon chain made of carbon and hydrogen atoms

Activity 4-3

- Your teacher will draw a diagram of the addition reaction of ethylene molecules on the board.
- Let each person in the class act as a hydrocarbon. Your torso is the carbon atom, your head is one hydrogen atom bonded to the carbon, and both feet together are another single hydrogen atom bonded to the carbon. Each hand is an electron that can bond with another electron on another molecule.
- All but 2 members of the class should find partners, face each other, and join hands. Each pair now represents an ethylene molecule, which has a double bond between two carbon atoms. Each pair of joined hands represents a bond made up of 2 electrons.
- Each of the two people who represent "unbonded" hydrocarbons should place one hand in a pocket and one hand out in front of them. The hand that is out represents a free electron that wants to bond with another electron. We will call these two hydrocarbons "radicals."
- The first radical should find an ethylene molecule and use the free electron hand to break a bond in the ethylene molecule, joining hands with one of the people in the molecule. This radical has just initiated the polymerization reaction. Now the molecule is left with an open electron.
- This open electron should approach another ethylene molecule, break a bond, and join hands with one of the people in the molecule. Continue this process, with free electrons breaking bonds and adding ethylene molecules to the chain.
- At some point, the other radical should come in and join hands with an open electron, ending the reaction. This radical has just terminated the reaction.

Condensation

The second type of polymerization reaction is the condensation reaction (Figure 4-8). Condensation occurs because when two monomers combine, they eliminate a small molecule, usually a water molecule.

Why do the two monomers eliminate a water molecule when they combine? The end of one monomer may have a hydrogen (H) with a partial positive charge. The end of the other monomer may have a hydroxyl (OH) group with a partial negative charge. The positive H and the negative OH can react with one another to form water, H_2O.

Meanwhile, the monomers are joining. As the positive H leaves its monomer, it leaves behind a negative charge (two electrons) on the end of the monomer. As the negative OH leaves its monomer, it takes two electrons and leaves a positive charge on the end of its monomer. The positive end of the one monomer and the negative end of the other monomer combine and form a bond linking the two monomers.

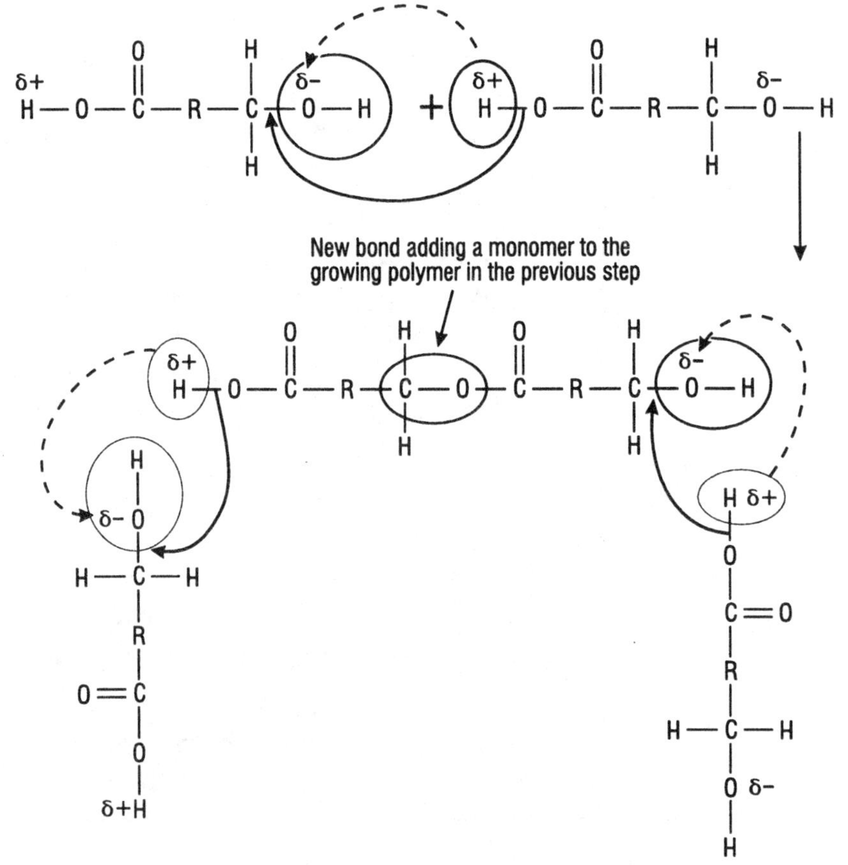

New bond adding a monomer to the growing polymer in the previous step

- R represents a hydrocarbon chain made of carbon and hydrogen atoms.
- The H+ ions follow the dashed arrows and bond with the partially negative OH to form water.
- The electrons left behind by the H+ follow the solid arrow and form a bond with the positive left behind by the OH⁻ that formed the water.
- The polymer chain can grow from either end.

Figure 4-8 Condensation polymerization reaction

The new, larger molecule now has a partially positive H and a partially negative OH on its ends. These will react with a partially negative OH or a partially positive H group on another monomer by the same reaction, eliminating a water molecule in the process. Monomers are added in this manner, one-by-one, to form a polymer. Each time a new monomer is added to the chain, a water molecule is eliminated.

The condensation reaction is much slower than the addition reaction, because it is not a chain reaction. To help the polymerization reaction, the water is usually removed as it is formed. If the water were not removed, the polymer chain would go through only a few reactions, making a very short polymer with a low molecular weight.

Activity 4-4

- Select twenty students to participate in the role playing of the condensation reaction. You will need four students for each of the following roles:
  - CO_2
  - CH_2
  - H
  - R
- Each student with the H role should hold hands with a student playing the CO_2 role. Each student playing the CO_2 role should also hold hands with a student playing the R role. Each student playing the R role should also hold hands with a student playing the CH_2 role. Each student playing the CH_2 role should also hold hands with a student playing an OH role. The arrangement should be

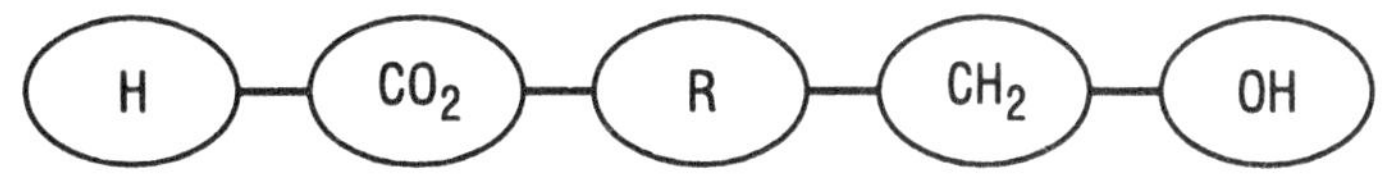

- The H on one molecule should approach the OH on another. The H should hold hands with the OH, forming a water molecule. The CO_2 and CH_2 should hold hands, forming the polymer molecule. The H end of the polymer molecule should repeat the process, and the H end of the unreacted monomer should also go through the same process. This gives a four-unit polymer and three water molecules.

What Is the Molecular Structure of Polymers?

Polymers are large molecules built up from smaller molecules called monomers. The polymer materials that have been traditionally used are wood, rubber, **cellulose** (sĕl′yə-lōs′) fibers, and protein fibers. These are all biological polymers that are found in nature. Some modern polymers include vinyl, polystyrene, and polyurethane.

Polymers consist of long-chain molecules made up of many repeating units (mers) joined together. These units come from small molecules called monomers. As more monomers add to a polymer, the molecular weight (the sum of the atomic weights of the atoms in a molecule) of the polymer increases—the polymer molecule becomes heavier. Figure 4-9 demonstrates the increase of the molecular weight with the growth of the chain.

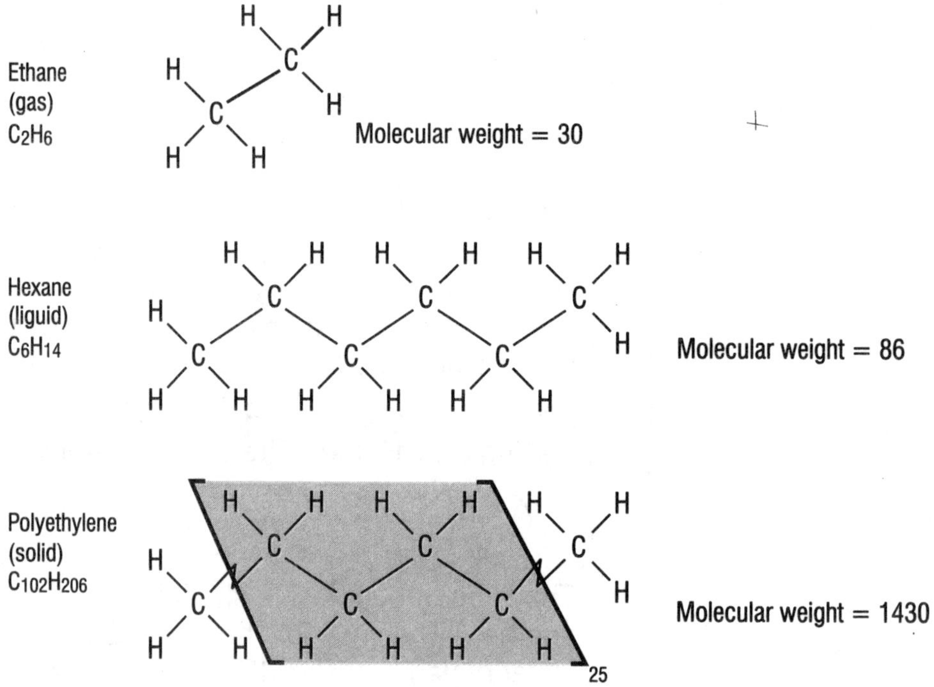

Ethane (gas) C_2H_6 — Molecular weight = 30

Hexane (liguid) C_6H_{14} — Molecular weight = 86

Polyethylene (solid) $C_{102}H_{206}$ — Molecular weight = 1430

Figure 4-9 Increase in molecular weight with increasing number of monomers

The properties of polymers depend on the way they bond, the length of the polymeric chains, and their molecular weight. Polymers can be classified according to their molecular structure.

Molecular Structure

The most important type of bonding in polymers is covalent bonding. There is some minor ionic bonding. Several types of covalent bonds are important in polymers. First are the carbon-carbon single bond and the carbon-hydrogen bond. These bonds constitute a great majority of the bonds in polymers. Other important bonds are the carbon-carbon double bonds, carbon-oxygen bonds, carbon-nitrogen bonds, and carbon-halogen (fluorine, chlorine, bromine, iodine, and so forth) bonds. While these bonds are fewer than the carbon-carbon and carbon-hydrogen bonds, they are very important. Many reactions in polymers occur at these other important bonds.

Polymer molecules may have a linear, branched, ladder, or network structure. Some polymers are linear, meaning that the structure is a long chain of atoms with side groups attached by covalent bonds. The polyethylene structure shown in Figure 4-6 is a linear polymer.

The molecules of a linear polymer (Figure 4-10a) do not form straight lines, but look more like a strand of cooked spaghetti. Many polyethylene molecules together resemble a bowl of spaghetti. The molecules are randomly intertwined, but the chains can be separated. When heated, the individual chains slip over each other, causing the polymer to flow.

Another type of polymer structure is a branched polymer (Figure 4-10b). The units or mers are linked into chains with branch points. At the branch points, side chains can grow. Essentially, a branched polymer consists of side branches of similar structure attached to a main chain.

As the name implies, a ladder polymer is formed when two linear polymers join in a regular sequence. Ladder polymers have a more rigid structure than linear polymers.

Network (or **cross-linked**) polymers (Figure 4-10a) are made of linear or branched polymers in which the chains and/or branches have joined. The cross links that join network polymers tend to form under stress, heat, and pressure. Whereas linear polymers are made of many molecules intertwined, a network polymer is one very large molecule. The more cross linking, the greater the strength and toughness of the polymer. Network structures have three-dimensional cross links. The cross bonds in cross-linked polymers prevent the individual chains from slipping.

Therefore, cross-linked polymers do not flow when they are heated.

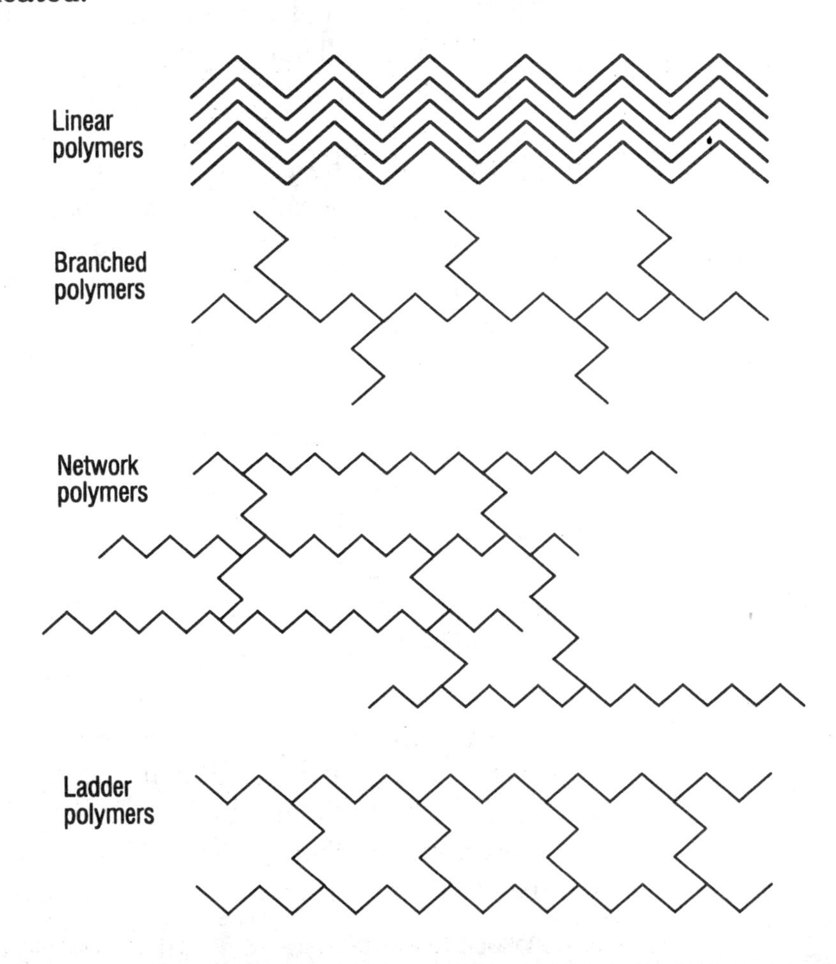

Figure 4-10 Different polymer structures

Activity 4-5

- Working in groups and using pipe cleaners, build models of the four kinds of polymer structures. Wherever the pipe cleaners are joined, make a small knot to represent a mer (a unit). Let the lengths of the pipe cleaners represent the bonds between mers.
- Compare the models of the different groups in your class. Decide if some work better than others as models for polymer structures.

Crystalline or Amorphous

You have learned that solid materials may be crystalline or amorphous. In crystalline materials, the molecules are joined in a stable repeating pattern. In amorphous materials, the molecules are not joined in such a stable or predictable way. How do the polymers just described—linear, branched, ladder, and network polymers—relate to the classifications of amorphous and crystalline solids?

As you might guess, most, but not all, network polymers are amorphous, without organization. Linear, branched, and ladder polymers are often, but not always, crystalline.

Crystallinity in polymers leads to stronger, stiffer materials with less impact resistance. In crystalline material, the polymer chains fold on themselves and pack together in an organized manner when solidifying. The resulting organized regions show the behavior of crystals. However, all crystalline polymers have amorphous regions between the crystalline regions. Figure 4-11 represents the two regions. Figure 4-11a shows the folded-chain polymer crystal. Figure 4-11b shows how the folded-chain polymer crystal regions are connected by amorphous regions in a type of polymer called a fringed micelle polymer crystal. The ordered part of the crystal is the micelle; the amorphous sections are the fringes.

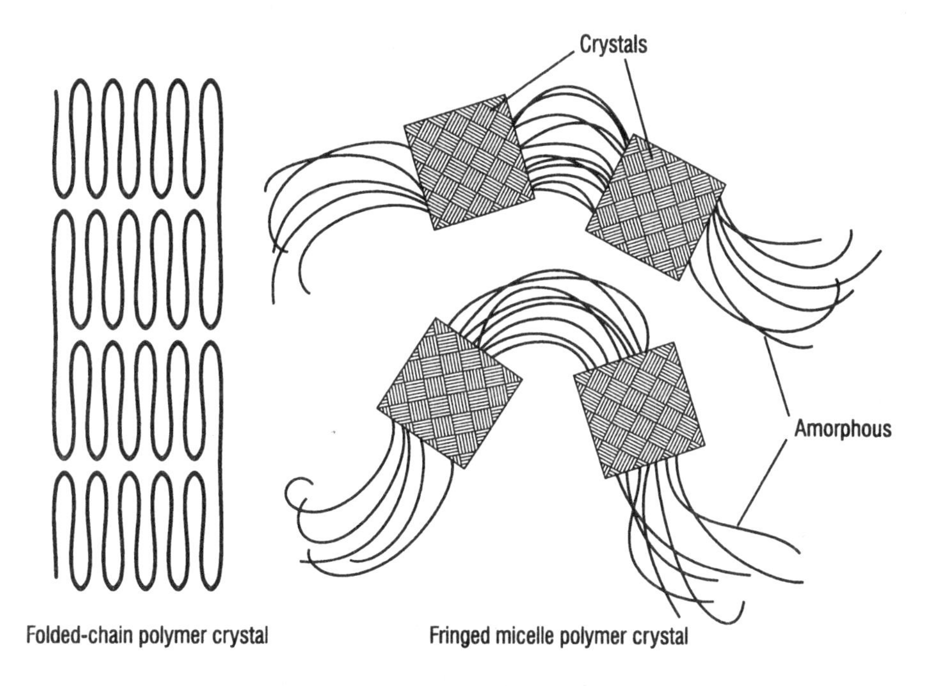

Figure 4-11 Types of polymer crystals

Crystalline polymers are more difficult to process, since they have higher melting temperatures and tend to shrink and warp more than amorphous polymers.

In industry, polymers are talked about as **thermoplastic** or **thermosetting**. Thermoplastic and thermosetting polymers are processed differently and have different properties.

Thermoplastic Polymers

Thermoplastics are polymers that melt when heated. When the polymer melts, it flows under pressure. These polymers can be repeatedly softened when heated and hardened when cooled. This property permits material to be reused and recycled. Most thermoplastics dissolve in specific solvents. Softening temperatures depend on the type and grade of thermoplastic. Most thermoplastics are amorphous linear polymers. They behave like independent but intertwined "spaghetti strands." Some thermoplastic polymers are crystalline. Table 4-1 lists some crystalline and amorphous thermoplastics.

Table 4-1. Some Crystalline and Amorphous Thermoplastics

Typical crystalline thermoplastic polymers	Typical amorphous thermoplastic polymers
acetal	polystyrene
nylon	ABS (acrylonitrile-butadiene-styrene)
polyethylene	polycarbonate
polypropylene	polyvinyl chloride
polyester	

Thermosets

Thermosets, or thermosetting polymers, are amorphous network polymers. These polymers chemically change during processing and become permanently solid. Prior to molding, a thermosetting polymer structure is very similar to that of a thermoplastic polymer. Cross linking of thermosetting polymers occurs during the molding process. Cross linking is the main difference between these two types of polymers. Cross linking makes the polymer insoluble to solvents and incapable of melting.

Elastomers

Elastomers (ĭ-lăs′tə-mərz), also called rubbers, are a special class of polymers that may be thermoplastic or thermosetting. Their structure is characterized by highly flexible, kinked sections that allow freedom of movement.

The term "elastomer" usually refers to a thermoplastic structure. The term "rubber" usually refers to a thermosetting structure. However, these terms are not always used in a precise way. Thermoplastic elastomers are highly elastic and flexible. Natural and synthetic rubbers such as latex, nitrile, and neoprene attain their properties through a process called **vulcanization** (vŭl′kə-nĭ-zā′shən). Vulcanization causes cross linking of the molecules, placing these rubbers in the thermosetting category.

What Are the Properties of Polymers?

In this section, we will see how the properties of plastics relate to the structures we have discussed. We'll find that there are predictable relationships, just as there are for metals and ceramics.

Among the thermoplastics, the simplest polymer is polyethylene—a carbon backbone with attached hydrogens. As you look down the chart through polypropylene to polytetrafluoroethylene (Teflon™), the nature of the molecules changes because the hydrogen atoms have been replaced with other atoms or molecules. These other atoms or molecules strengthen the polymers, and even alter the chemical reactivity. For example, Teflon has a low reactivity, which allows it to be used as a coating for pots and pans as well as for gaskets and columns used in the chemical industry.

When additional molecules are grafted onto the carbon backbone, as in the acrylics and polycarbonates, the polymer chain is stiffened. When polar groups are added, such as in nylon, hydrogen bonding further increases the strength. The cross linking that occurs with polyesters results in a rigidity that approaches that of thermosetting materials.

The main feature of thermosets is the cross-linked network structure that produces essentially one large molecule. The properties of the thermosets can be changed by adding fillers or plasticizers. A filler is a material added to a polymer to strengthen or modify its properties.
A plasticizer is a lower-molecular weight chemical added to a polymer to soften it.

Some Important Properties of Polymers

Manufacturers select materials for their products based on the properties of those materials. Polymers make up a diverse group of materials with many different properties. Here we will discuss just a few of the properties that may be important in choosing a polymer for a particular project.

Tensile Strength

The tensile strength of a material is its resistance to that force tending to pull the material apart. In applications where strength is important, manufacturers pick a material based on its tensile strength. In the thermoplastics group, the highest tensile strengths are found in the polymers with stiffened carbon chains, such as the nylons and acetals. The thermosets generally have higher tensile strengths than the thermoplastics. Why do you think this is so?

Percent Elongation

Percent elongation is the change in the length of a material when it is subjected to a load, as expressed in Equation 4-1. The term “strain” is the same measurement expressed as a ratio (Equation 4-2) instead of a percentage.

$$\%\ \text{elongation} = \frac{\text{Final length} - \text{Initial length}}{\text{Initial length}} \times 100\%$$

Equation 4-1

$$\text{Strain} = \frac{\text{Final length} - \text{Initial length}}{\text{Initial length}}$$

Equation 4-2

The property of percent elongation is useful to know when materials are expected to undergo stretching. The higher the percent elongation, the more ductile the material. A ductile material is one that can be plastically deformed by elongation. Conversely, the lower the percent elongation, the more brittle the material.

For example, among the thermoplastics, there is a big difference between polyethylene and polystyrene. The large styrene groups on the carbon chain of the polystyrene prevent the molecules from easily slipping over each other. Therefore, polystyrene has a lower percent elongation than polyethylene. Which of the two materials would you expect to break more easily? Thermosetting polymers, with their network structure, are unable to uncoil. Therefore, most of the thermosets have very little or no elongation.

Hardness

As with metals, higher hardness roughly indicates higher strength. However, hardness is not indicative of wear resistance. For example, the thermoplastic polymer acetal has an outstanding wear resistance but a low hardness value.

Modulus of Elasticity

Modulus (mŏj′ə-ləs) **of elasticity** is a measure of a material's deformation over a range of loads or stresses (the material's ability to deflect). The modulus of elasticity, E, is defined as the stress a material undergoes divided by the strain the material demonstrates under that stress. Strain is expressed in Equation 4-2. Stress is expressed in Equation 4-3.

$$\text{Stress} = \frac{\text{Applied force}}{\text{Cross-sectional area of specimen}}$$

Equation 4-3

$$E = \frac{\text{Stress}}{\text{Strain}}$$

Equation 4-4

Polymers have a much lower modulus of elasticity than do metals and ceramics. Therefore, a plastic may not be able to be substituted for a metal or ceramic if the part is subject to high deflection.

Activity 4-6

- Get the following from your teacher:
  - rubber band
  - micrometer
  - spring scale (0 to 18 oz)
  - ruler
- Use the micrometer to measure the width (w) of the rubber band. Record this value in your *ABC* notebook.
- Use the micrometer to measure the thickness (t) of the rubber band. Record this value in your *ABC* notebook.
- Use the ruler to measure the length of the rubber band. Record this value in your *ABC* notebook.
- Loop the rubber band over the zero end of the ruler and the hook of the spring scale. Pull on the spring scale to stretch the rubber band until there is a 3-ounce force on the spring scale. Record the 3-ounce force and the stretched length of the rubber band in a table in your *ABC* notebook.
- Repeat step 4, applying the following forces to the rubber band and recording the force and length values in the table in your *ABC* notebook:
  - 5 ounce
  - 8 ounce
  - 12 ounce
  - 15 ounce
- Multiply w × t × 2 to get the cross-sectional area of the rubber band. (The 2 is included because two sides of the rubber band are stretching.)
- Use Equation 4-3 to calculate the stress on the rubber band at each applied force.
- Use Equation 4-2 to calculate the strain on the rubber band at each applied force.
- Use Equation 4-4 to calculate the modulus of elasticity at each applied force.
- Record your findings in your *ABC* notebook for use in the Unit Wrap-Up Activity.

Testing Polymers

CAREER PROFILE: POLYMER TESTING TECHNICIAN

John S. is a polymer testing technician for a plastic company. The company makes raw plastic—or, as they usually refer to it, virgin plastic—plastic that has never before been used to make a product; it contains no reground or recycled materials.

At this particular company, raw chemicals are purchased from suppliers. They are put through reactors and made into a polymer—either polyethylene or polypropylene. Then the plastic is extruded into pellets. [To extrude something is to force it out through a shaped pattern—in this case, a pellet shape.]

As the polymer is made, certain materials are added to it to make it more stable or give it other characteristics. For example, **antioxidants** (ăn′tē-ŏk′sĭ-dənts) are added to protect against degradation from heat and stress during processing; UV stabilizers protect against degradation from the sun's ultraviolet light; antistatic additives prevent static. After processing, the pellets are sold to customers—other companies that may make milk bottles, plastic cups, gas tanks, or other products out of the virgin plastic.

John explains his role in the company. "My specific role is to do additive testing. I extract the additive from a given polymer and test it. The test methods that I use most often are **high-performance liquid chromatography** (krō′mə-tŏg′rə-fē) (HPLC) and **gas chromatography** (GC)." [See this section for an explanation of HPLC.]

"The testing I do now is primarily for a research team in the company, but I used to work in the quality assurance laboratory. There we would take a sample of the powdered plastic right after it had gone through the reactor. We did several tests on the sample, including a melt index in which we measured the flow at certain temperatures. We also did a density test and tested for impurities in the polymer."

To prepare for his job, John took chemistry in high school and went through a two-year chemical technology program at a technical college. "That was a good preparation for me," he says. "Even through a lot of what I know now I learned on the job, the program gave me a background. The company does expect an incoming technician to know something. If they talk to you about a benzene ring for example, they don't want you to say, 'what's a benzene ring?' They want you to picture a ring of six carbon atoms."

John has been at the plastics company for almost five years. He works with a team of chemists and chemical technicians. "I work with the additive chemist—she comes in and works with me on a day-to-day basis. I also do work with the analytical lab manager. Everything is based on teamwork," he explains. "If you can't work as part of a team. . . well, you may be in trouble. Our jobs depend on how well the whole team, and the company, perform."

Activity 4-7

Imagine for a moment that you are a manufacturer of certain stereo components that are made out of a polymer. These components are rather expensive to form and are used in a high-priced line of stereo equipment. You have chosen a certain polymer to make these components because of its properties. You know that it has certain strength, hardness, and so on that are absolutely necessary for your product. When the shipment of polymer resins comes in to your factory, how will you handle it? Discuss these alternatives as a class.

- Direct the entire shipment directly to the production area and rely on the test information provided by the supplier.
- Send a portion of the shipment to production to be tried out to verify its quality and hold the rest until that production run is completed.
- Have the shipment sampled and tested, holding the shipment in the receiving area until the test results are confirmed.

Polymers are tested to verify their properties by the manufacturers purchasing them. The makers of polymers test them to make sure that they will meet the specifications of buyers.

Earlier, you read about some of the physical properties that are often tested, such as hardness or percent elongation. Chemical properties are also tested. A few of the chemical tests used to verify these chemical properties are high-performance liquid chromatography (HPLC), a solution viscosity test, and the melt index test.

High-Performance Liquid Chromatography

High-performance liquid chromatography is a technique that separates a mixture based on the molecular weight or the molecular weight distribution of the component.

In HPLC, a sample of a material is dissolved in a liquid solvent; the liquid is referred to as the mobile phase because it moves. It is passed through a stationary column in which there is a material that adsorbs (takes up) the liquid phase as it moves through the column. Because different compounds in the sample adsorb to the stationary phase to different degrees, they go through the column at different rates. Thus they emerge from the column separately, and each one is detected by a light source and its chemical content is indicated, based on the light detected. Different materials absorb different wavelengths of light, so the particular wavelengths absorbed indicate the chemical identity of the material. The amount of each compound within a material may also be indicated, depending on the equipment.

Solution Viscosity Test

Solution viscosity is a measure of the size of the polymer molecule and is therefore related to the molecular weight of the polymer. To measure solution viscosity, the polymer is dissolved in a solvent and the time it takes this solution to flow through a capillary tube is measured (t). This time is compared to the time it takes the solvent alone (without the polymer) to go through the same tube (t_0). The concentration of the solution is also considered; however, in simple terms solution viscosity can be expressed as in Equation 4-5.

$$\text{Solution viscosity} = \frac{t}{t_0}$$

Equation 4-5

Most processes used to shape thermoplastics involve melting the polymer. The melt flow index (MFI or MFR—melt flow rate) measures how well the polymer flows when heated to a stated temperature. In this test, a specified weight is used to force the heated polymer sample through a hole in a standard die. For example, MFI (210, 5) = 4.5 means that 4.5 grams of the polymer are forced through the die in 10 minutes by a 5-kilogram weight at 210°C. End-product manufacturers use this number to aid in determining the setup conditions of their processing equipment, such as extruders or injection molders.

How Are Polymers Made into Products?

CAREER PROFILE: POLYMERS MACHINIST

Ken C. is a machinist for a company that makes products from polymer materials. Ken's company typically makes small batches of any given product, customizing it to meet the buyer's needs. Their products range from the applicator sticks used in eye makeup pencils to the O-ring seals used on rocket engines.

Ken explains that the raw materials for the company's products usually come from larger petrochemical companies. "Our raw materials are not ordinary plastics," he explains. "They are what we call 'high polymers' such as Teflon, or a relatively new material called polyetheretherketone (PEEK).

"The materials come in looking like rabbit food, little bunny pellets, or sometimes, like soap flakes or soap powder. Through the use of heat and pressure, we form that polymer into a finished article—some kind of shape. Here the heat and pressure processes are extrusion and injection molding, primarily. Sometimes the shape is complete and ready to go; other products require further machining. The expertise of our company is in the area of machining polymers.

"I started learning metals machining in high school, and I worked in a metal fabrication shop after I graduated. I came here nine years ago. It took me quite a while to gain the skill of machining polymers. With some of these materials, well, it's like machining solidified jello. It's almost rubbery—not like metal at all."

Ken uses a lathe that is similar to a metal lathe. It differs in that it has a hollow spindle into which the polymer material is fed. It also has an entirely different set of cutting tips from those used in metal machining.

"When I train new machinists—and that's a big part of my job now—I try to teach them something about the properties of the materials. For example, Teflon is very slick and very abrasive to metals. So the tips used with Teflon tend to wear down quickly. If you allow them to wear unevenly, the product will come out oversized or undersized. You have to get a feel for it; after a while you learn what kinds of in-process adjustments are going to be necessary. You learn not to let the tip get too hot, or you'll be melting polymer instead of cutting it. You learn not to cut so much that you're pushing the material instead of cutting it. There's a lot to learn."

As you might expect, Ken's skills are highly valued in his company. The plant manager, a chemical engineer, emphasizes that it isn't easy to find machinists who can work with plastic. "It's a scary experience from the outset for a metal machinist to machine this plastic material. We need more people like Ken."

Some polymer processing techniques are similar to those used to form metals and ceramics. Manufacturers can buy sheets, rods, or tubes made of some polymers and use them to make plastic parts. However, often a chemical mixture of polymers is compounded or mixed in the equipment as a part or product is formed. In these cases, the polymers used in the compounding are selected based on the properties required for the product.

Five basic processes are used to form polymer products or parts. These are **injection molding**, **compression molding**, **transfer molding**, **blow molding**, and **extrusion**. Compression molding and transfer molding are used mainly for thermosetting resins. Injection molding, extrusion, and blow molding are used mainly for thermoplastic polymers.

Injection Molding

Injection molding is the most common process used to produce plastic products. Injection molding involves four steps:

1. Polymer, in the form of powder or pellets, is usually heated to a liquid state.
2. Under pressure, liquid polymer is forced into a mold of the desired shape. The polymer fills the mold through an opening called the sprue. Runners carry the polymer to the mold cavities. Gates control the flow of material into the cavities.
3. The material is held under pressure in the mold until the polymer solidifies.
4. The mold is opened and the part is removed.

Figure 4-12a is a schematic drawing of a basic injection molder. Figure 4-12b is a cross section of a mold.

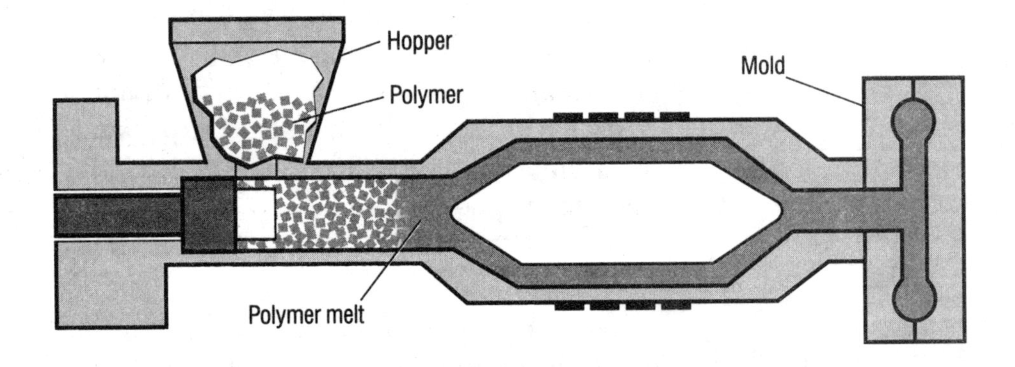

Figure 4-12 Injection mold

Injection molding is a rapid process that produces little waste and can be highly automated. Plastic parts can be produced in large quantities (for simple parts, one mold can consist of many cavities to produce several parts at once). In most cases, the products coming out of the mold are finished parts. Only the sprue and runner pieces must be removed. Model airplane parts, combs, drink stirrers, and toothbrush bases are examples of injection-molded parts.

Compression Molding

Compression molding was one of the first processes developed to mold plastics. Figure 4-13 is a compression mold. Raw materials in the form of preformed blanks, powders, or pellets are placed in the bottom section of a heated mold (or die). The other half of the mold is lowered and pressure is applied. The material softens under heat and pressure and flows to fill the cavities of the mold. Any excess material is squeezed out of the mold. The mold is opened and the part is removed using ejector pins or plugs. For thermoplastics, the mold must be cooled before injection—otherwise the part will lose its shape. Thermosets can be ejected while they are hot, after curing is complete.

Figure 4-13 Compression mold

Two advantages to compression molding are 1) the material moves a short distance into the mold and 2) the material does not have to flow through runners and gates. However, the time needed to heat and cure the plastic makes compression molding a slow process. Generally, only one part is made from each mold during the compression molding process.

Transfer Molding

Transfer molding is a modified process of compression molding. It eliminates the turbulence and uneven polymer flow caused by compression molding. Like compression molding, transfer molding is used mainly for thermosetting plastics.

A partially polymerized material is placed in a heated chamber. At this stage, the material will flow but it will not cross link. A plunger forces the hot, liquid polymer into the mold cavities. The material flows through sprues, runners, and gates. The temperature and pressure inside the mold are higher than in the heated chamber. Thus, cross linking occurs in the mold. After the plastic has cured and hardened, the mold is opened and the part is removed.

Complex parts with varying thickness can be accurately produced using transfer molding. However, mold costs are very high and a significant amount of scrap material is generated in the sprue and runners.

Extrusion

Extrusion is used to make shapes of products with constant cross sections, such as pipes and rods. Extrusion forces molten polymer through an opening, called a die, to produce the final shape. The process involves these four steps:

- Compounding (mixing) and pelletizing the plastic (coloring agents and other additives are mixed with the polymer at this time)
- Heating the material to the proper plasticity
- Forcing the material through the die
- Cooling the material

An extruder contains a hopper to feed polymer and any additives into the machine, a barrel containing a continuous feed screw, a heating element, and a die holder. Figure 4-14 is a drawing of a typical extruder.

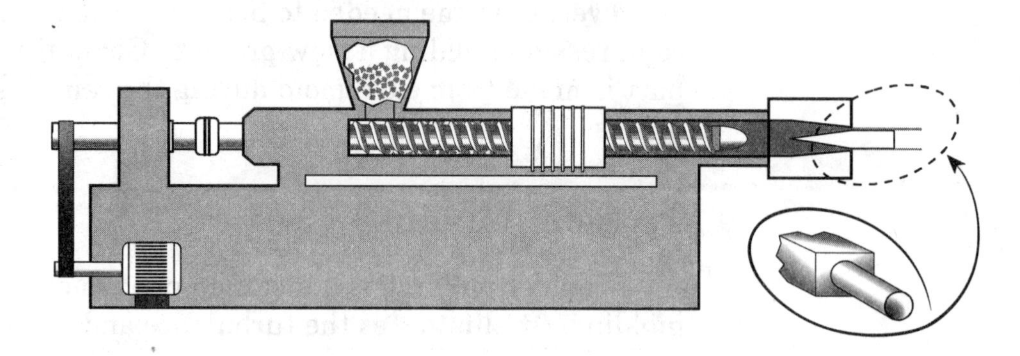

Figure 4-14
Extruder

In addition to rods and pipes, extruders are used as compounders or mixers. The output from an extruder compounder is chopped or pelletized to form the feedstock for another process such as injection molding. An adapter can be added at the end of the extruder that blows air through the orifice while the hot polymer is exiting. Plastic bags and films are made using this adapter.

Blow Molding

The manufacture of plastic bottles such as those used for milk involves a process called blow molding. Hot water bottles, pails, globe light fixtures, gasoline tanks, and 55-gallon drums are also produced using this process. In blow molding, a softened plastic tube is inflated to fill a mold cavity. Only a small opening is needed to blow in the air. Therefore, rapid manufacture of products with a solid shell is possible.

Activity 4-8

- Divide the class into five groups. Let each group contact a different petrochemical company. (Your teacher will provide names and addresses.) Inquire about the following:
  - Materials used in the processing of oil and gas to make polymers
  - Worker safety and health issues related to the use of materials and/or processes in polymer production
  - Company safety and health measures, including recent health and safety statistics
- As a class, compare and contrast the material received from the different companies. Discuss the following:
  - What appear to be the most critical health and safety issues in polymer processing?
  - Based on their literature, how do the different companies work to prevent or respond to health and safety issues?
  - Based on their literature, in which companies do health and safety seem to be a high priority?
  - How would you conduct further research to find out about the health and safety practices and records of these companies?

Looking Back

Polymers are made by linking molecules called monomers. In a reaction called the addition reaction, monomers activated by an initiator link to form very long chains. In another process of polymerization, water molecules are formed—a condensation reaction.

The molecular structure of polymers helps to explain why they have the properties they do and why they are processed in certain ways. Polymers may be linear, branched, or ladder polymers, which are often crystalline and thermoplastic (those that melt when heated). Network polymers are polymers in which many different branches are cross linked; as a result they are insoluble and incapable melting. They are amorphous—their cross linking does not follow a predictable course. They are also thermosetting—during processing, they become permanently solid.

Like other materials, polymers are selected for use in products on the basis of their properties. Some of these properties are tensile strength, percent elongation, hardness, and modulus of elasticity. Polymers are tested for these and other properties by the companies that make them and the companies that buy them. High-performance liquid chromatography and solution viscosity are two tests often used to verify the chemical composition and chemical properties of polymers.

Polymers may be processed to make products by injection molding, compression molding, transfer molding, extrusion, and blow molding. All of these methods involve setting up proper conditions of heat and pressure to produce a part with the desired shape.

Further Discussion

- Explain why you think network polymers are thermosets and why linear, branched, and ladder polymers are primarily thermoplastic. Which of the forming methods described seems best suited to each type of polymer?

- What qualities do you think might be required for each of the two jobs described in this subunit: polymer plant operator and polymer test technician? Which do you think would require more training? Which would be the more interesting, in your opinion?

Activities by Occupational Area

General

How Are Monomers Made?

- Using library resources or the worldwide web, research the monomer ethylene. Find out how it is made from petroleum and what kinds of polymers are made from it. Write a short report in your *ABC* notebook.

Agriscience

Soil Amendments

- Contact one or more plant nurseries in your area and find out if polymer materials are used in soil to improve aeration, provide mulch, or otherwise promote plant growth and health. How do such polymers compare with naturally occurring materials used for such functions?

Health Occupations

Air Pollutants from Polymer Plants

- Contact environmental agencies and organizations as well as chemical manufacturing groups to find out what air pollution problems are associated with polymer production. Identify the problems and proposed solutions. Identify which companies have made the greatest progress so far in solving these problems. Report your information to your class.

Family and Consumer Science

Recycling of Plastics

- Assume you are a conservation consultant for a city sanitation department. The city is interested in expanding its recycling program to include plastic materials. Develop a report for the city council that explains 1) how plastics are classified for recycling and what the classifications mean, 2) where recycled plastics are taken, and 3) how they are processed. Include your report in your *ABC* notebook.

Industrial Technology

Natural and Synthetic Rubber

- Research the use of natural and synthetic rubber, especially the impact of the development of synthetic rubber on the outcome of World War II.

LAB 7

SEPARATION OF A MIXTURE BY DISTILLATION

PREVIEW

Introduction

You might have seen old movies of oil strikes. The land is always flat as a pancake for as far as you can see. The oil drilling has been going on for hours or days. The clock is ticking while the dirt farmers who staked their last dime on the claim look on with desperate faces, at the very brink of admitting defeat. Suddenly, at the eleventh hour, the gusher shoots out of the flat land, black as ink, spurting thousands of feet upward before raining down on the grateful, a-minute-ago-poor-and-now-filthy-rich farmers below.

That's the end of the first part of the movie; next comes the part where you find out which members of the family will be morally corrupted by the "black gold" and which ones will not. But did you ever wonder what happens to the gusher? After they cap it, set up the oil rig, pump it, and transport it by rail or sea or highway or pipeline—what then?

Crude oil, as petroleum pumped from underground is called, goes through several processes before it can be put to most of its uses. The most important of these is called the distillation process.

During distillation, the crude oil is heated and pumped into a fractionating tower. Substances with the lowest boiling points rise to the top of the tower and are drawn out as gases. Other substances condense and fall into trays or platforms, which are set at different heights in the column and "capture" the condensation according to its boiling point (see Figure L7-1). Some substances stay in the liquid state and are drained off at the bottom of the column.

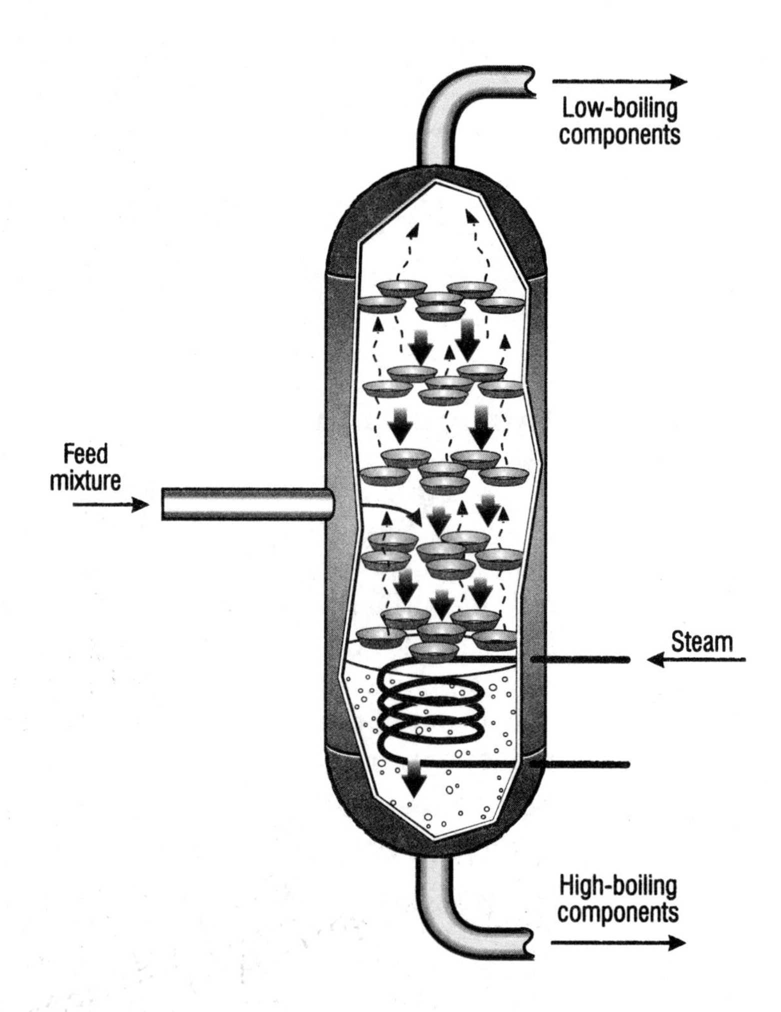

Figure L7-1
Distillation column

Purpose

In this lab, you will determine the relative boiling points of two liquids and a mixture of the liquids.

Lab Objectives

When you've finished this lab, you will be able to—

- Determine the approximate boiling point of a liquid.
- Separate a mixture of liquids based on their boiling points.

Lab Skills

You will use these skills to complete this lab—

- Measure a liquid volume with a graduated cylinder.

- Read a thermometer to determine an approximate boiling point of a liquid.

Materials and Equipment Needed

acetone
mixture, 50% acetone and 50% water by volume
water
3 250-ml Erlenmeyer flasks
Thermometer, –20 to 120°C
6 boiling chips
hot plate
ring stand
thermometer clamp
hot pad
stopwatch, clock, or timer
safety goggles
lab apron

Pre-Lab Discussion

Boiling a liquid seems simple enough. You simply heat the liquid until it begins to boil. But what is really happening? At any given temperature, a liquid substance has a vapor pressure—the pressure of the vapor above the liquid. As the temperature of the substance increases, its vapor pressure also increases. When the vapor pressure of the substance is equal to the atmospheric pressure, the substance begins to boil. At this point, the bubbles of vapor can form, move to the surface of the liquid, and escape.

What happens if the atmospheric pressure increases? In this case, the vapor pressure would not be enough to overcome the atmospheric pressure. Before the substance could boil, the temperature would have to increase, giving a higher vapor pressure. If the atmospheric pressure decreased, the boiling point—temperature needed to boil the substance—would decrease. For this reason, the boiling point of a substance is always reported at one atmosphere of pressure.

If two substances are mixed together, they can be separated based on their vapor pressures at a given temperature or their boiling points. If the substances are heated in open vessels, they will boil near their boiling points. At any given temperature, vapors of both components will be liberated from the liquid based on the vapor pressure of that component. A low-boiling component will have a higher vapor pressure at any given temperature than a higher-boiling component. As the mixture is heated to the boiling point of the lower-boiling component, the mixture will begin to boil and the temperature of the mixture will remain constant until the lower-boiling component is completely removed from the mixture.

This difference in vapor pressures and boiling points is the basis for the separation of components of a mixture by distillation.

Safety Precautions

- Do not use open flames during this lab. Acetone is very flammable.
- Do this experiment in a fume hood or well-ventilated area.
- Use hot pads when handling hot flasks.

LAB PROCEDURE

Method

Put on your lab apron and goggles.

1. Pour 75 ml of acetone into a 250-ml Erlenmeyer flask. Label this flask "acetone."
2. Add two boiling chips to the Erlenmeyer flask, and place the flask onto the hot plate.
3. Position the ring stand with a thermometer clamp, so a thermometer can be positioned in the flask approximately 1 cm above the surface of the liquid in the flask, as shown in Figure L7-2.

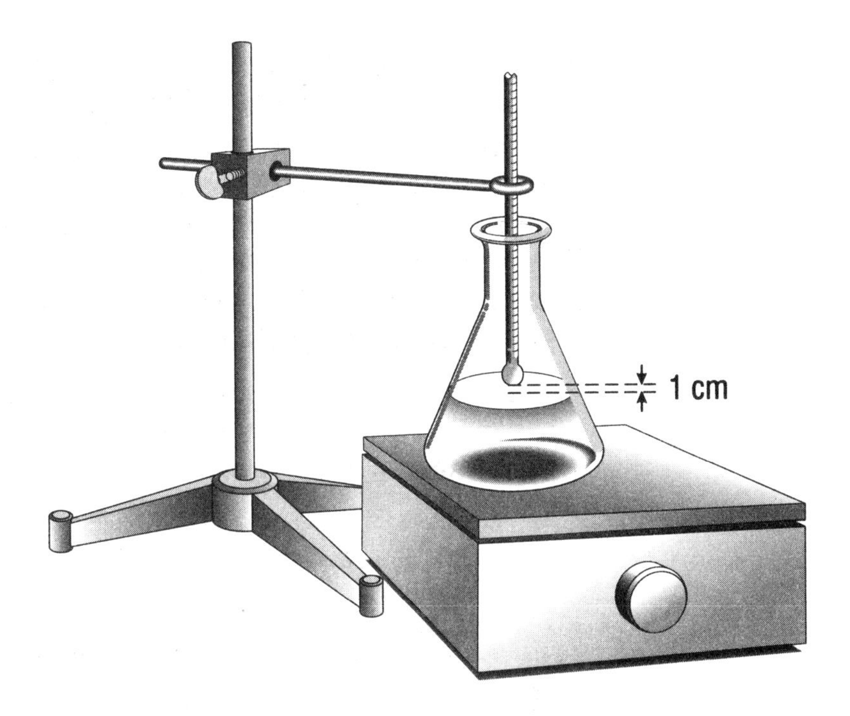

Figure L7-2 Setup of thermometer and flask

4. Turn on the hot plate at a medium power and observe the temperature in the flask.
5. When the flask begins to boil, allow it to boil for two minutes and then record the temperature of the vapor in the flask in your *ABC* notebook.
6. After the flask boils and the temperature has been recorded, turn off the hot plate and carefully remove the flask from the hot plate with the hot pad.
7. Pour 75 ml of distilled water into a second 250-ml flask. Label this flask "water." Repeat steps 2 through 6 for the "water" flask.
8. Pour 75 ml of the 50% acetone- and 50% water-by-volume mixture into a third 250-ml flask. Label this flask "mixture." Repeat steps 2 through 6 for the "mixture" flask.

Cleanup Instructions

- Empty the liquids into the sink and run water into the sink for about one minute.
- Wash and dry the glassware and return it to its proper storage location.
- Return the hot pad, hot plate, ring stand, and thermometer clamp to their proper storage locations.

Calculation

No calculations are required for this lab.

WRAP-UP

Conclusions

1. What was the temperature of the acetone vapor two minutes after the acetone flask began to boil?
2. What was the temperature of the water vapor two minutes after the water flask began to boil?
3. What was the temperature of the mixture vapor two minutes after the mixture flask began to boil?
4. Based on these temperatures, what do you think the vapor in the mixture flask is?

Challenge

5. Describe a method to separate acetone and water based on their boiling points.

LAB 8

MAKING NYLON 66

PREVIEW

Introduction

Cindy N. is a materials engineer for a polymer processing plant. Her supervisor has given her a project to troubleshoot. For some reason, the nylon fibers being produced do not have the required strength. This problem has been occurring for about three weeks.

Cindy has drawn up a list of possible causes:

1. The reaction to make the nylon has changed—
   - Are the raw materials being fed in the same proportions as before?
   - Have the raw materials changed?
2. The fiber-drawing process has changed—
   - Is the fiber drawn at the same speed and tension?
   - Has the drawing temperature changed?

After talking to the process operator and the industrial chemist, Cindy has ruled out the possibility of the raw materials being fed differently. She has called the raw-materials supplier. The technical representative has assured her that the raw materials have not changed. This has been confirmed by the results from the samples she sent to the lab.

The drawing temperature is monitored by thermocouples that are calibrated every day, so Cindy has been able to eliminate that factor from her list. Just as Cindy is beginning to feel stumped, an operator happens to mention to her that the tension gauge has been giving the operator a headache. The gauge keeps bouncing all over the place. Cindy calls in a mechanic to look at the gauge. He tells her that the gauge is over 20 years old and is no longer reliable. Next Cindy consults a mechanical engineer, who recommends a type of gauge that controls the tension better than any other on the market. Cindy follows up by ordering the gauge. Within two days, Cindy has solved the problem, but she hasn't done it alone. How many members of the production team (including outside vendors) has Cindy consulted?

Purpose

In this lab, you will make a nylon polymer using a condensation reaction.

Lab Objective

When you've finished this lab, you will be able to—

- Understand how nylon polymers are made.

Lab Skills

You will use these skills to complete this lab—

- Accurately measure the volumes of solutions with a graduated cylinder.
- Read a Material Safety Data Sheet (MSDS).

Materials and Equipment Needed

0.5-M hexamethylenediamine in 0.5-M NaOH
0.25-M adipoyl chloride
acetone
washbottle
250-ml beaker
large watch glass
2 10-ml graduated cylinders
18-mm × 150-mm test tube
gloves
lab apron
goggles
copper wire

Pre-Lab Discussion

The name "nylon" refers to synthetic polyamids. Nylon 66 is formed by the condensation reaction of hexamethylenediamine and adipoyl chloride—

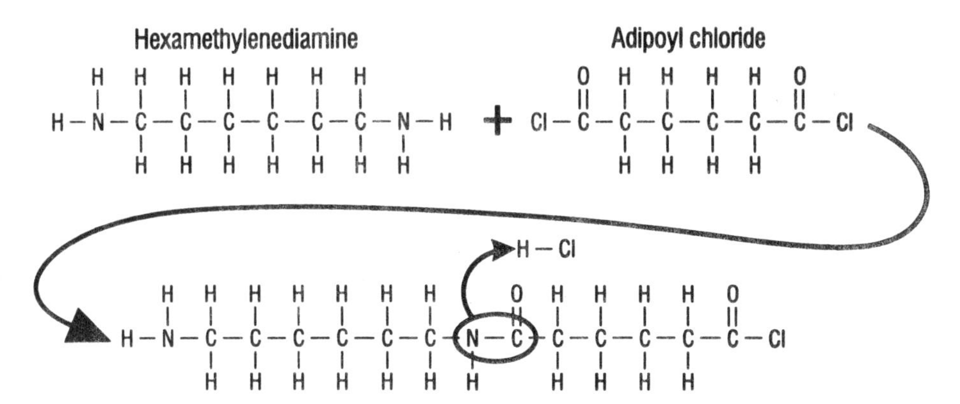

First step product of condensation reaction

This reaction is a condensation reaction. It takes place at low temperatures (that is, room temperature) and does not depend on an exact ratio in the reacting mixture.

The choice of solvents (hexane and water) in this lab is very important. These solvents do not dissolve in each other. Therefore, the solutions do not mix either. Instead, they form an interface (surface between the two solutions). This interface is the only place the reacting species can come into contact.

Safety Precautions

- Before beginning this lab, get copies of the Material Safety Data Sheets (MSDS) for the hexamethylenediamine and the adipoyl chloride. Read this material and observe all cautionary statements.
- Do this lab in a fume hood or a well-ventilated area.
- Do not have open flames in the lab when working with acetone.

LAB PROCEDURE

Method

Put on your lab apron, gloves, and goggles.

1. Bend a piece of the copper wire into a hook. The shaft should be long enough to reach the bottom of the 50-ml beaker.

2. Using the 10-ml graduated cylinder, measure 3 ml of the hexamethylenediamine solution. Pour this solution into the 50-ml beaker.
3. Using the other clean, 10-ml graduated cylinder, measure 3 ml of the adipoyl chloride solution. Carefully pour this solution onto the surface of the hexamethylenediamine solution. Observe what happens, and write your observations in your *ABC* notebook.
4. Insert the wire hook into the beaker and catch the film on the hook.
5. Slowly pull the hook from the beaker, and observe what happens in the beaker as the film is removed. Record your observations in your *ABC* notebook.
6. Continue to withdraw the film. When the thread breaks, place it on a watch glass and wash it with water, and then wash it with acetone. (As an alternative to placing the threads on a watch glass as they break, you can wind the thread around an 18-mm × 150-mm test tube. Be sure to wash the thread with water and then acetone.)
7. Carefully examine your polymer and write your observations in your *ABC* notebook.

Cleanup Instructions

- Any reaction mixture remaining in the beaker should be thoroughly mixed to produce nylon. Wash the solid nylon with water and then acetone before discarding the solid nylon.
- Clean all glassware and return the clean, dry glassware to its proper storage location.

Calculation

No calculations are needed for this lab.

WRAP-UP

Conclusions

1. What part does the interface between the two solutions play in the nylon 66 condensation reaction?
2. How would the reaction in this lab have been different if different solvents that were soluble in each other had been used?

Challenge

3. Is the nylon thread that you made as strong and flexible as the nylon used in clothing such as pantyhose? Compare your product to a sample of nylon from your teacher.

LAB 9

CROSS LINKING A POLYMER

PREVIEW

Introduction

Sarah S. is an area supervisor for a tire manufacturer. Today, she is giving an orientation session for new employees. She is explaining how tires are made to the new employees. "Welcome to your first day on the job. I'm here to tell you, briefly, how we make tires. You will learn more of the specifics when you get to your area.

"Basically, we take a rubber, such as isoprene, and add sulfur to cross link the polymer chains. I'll explain the chemical reactions.

"First, we take the isoprene monomer and polymerize it.

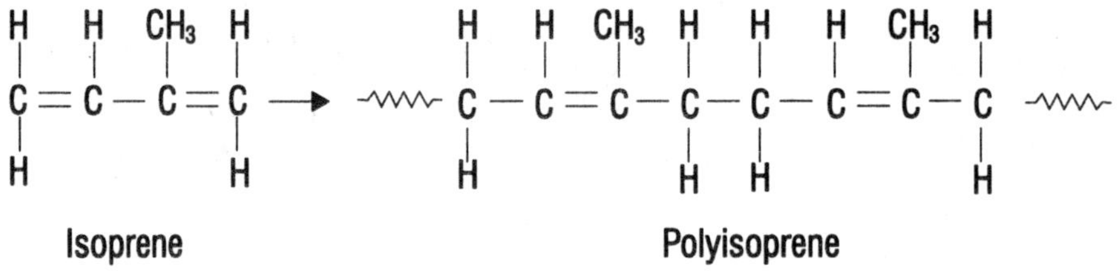

Isoprene → Polyisoprene

Figure L9-1 Isoprene polymerization

"When many polyisoprene chains are floating around, we add sulfur to cross link the polyisoprene chains.

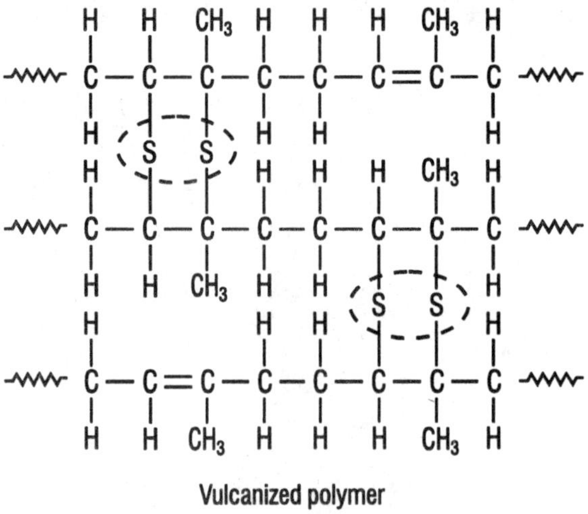

Vulcanized polymer

Figure L9-2 Cross-linking

"Cross linking the polymer in latex by adding sulfur is called vulcanization. Cross linking increases the strength, hardness, and toughness of the rubber. As the sulfur content increases, the rubber changes from a gummy material to a tough, elastic substance such as that used in tires. Further increases of the sulfur make an even harder rubber, like that used in flexible combs or battery cases. This rubber wouldn't make good tires, so the amount of sulfur must be controlled carefully.

"I hope I have been able to give you a brief explanation of the tire-making process. I'm available at any time when you have questions."

Purpose

In this lab, you will create a cross-linked polymer.

Lab Objectives

When you've finished this lab, you will be able to—

- Demonstrate a chemical process.
- Understand the concept of cross-linking.
- Understand the concept of hydrogen bonding.

Lab Skills

You will use these skills to complete this lab—

- Accurately measure solutions using a graduated cylinder.
- Read a Material Safety Data Sheet (MSDS).

Materials and Equipment Needed

4% aqueous solution of 99-100% hydrolyzed polyvinyl alcohol
4% solution of borax ($Na_2B_4O_7 \cdot 10H_2O$)
2 clear plastic cups
wood splints (popsicle sticks)
food coloring
water-soluble pen
table salt
lab apron
safety goggles

Polyvinyl alcohol is a water soluble polymer. The chemical formula is—

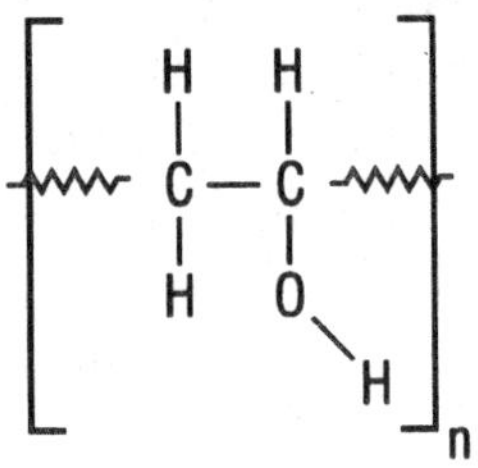

$$\left[\sim\!\!\sim \underset{\substack{| \\ H}}{\overset{\substack{H \\ |}}{C}} - \underset{\substack{| \\ O \\ \diagdown \\ H}}{\overset{\substack{H \\ |}}{C}} \sim\!\!\sim \right]_n$$

Pre-Lab Discussion

Borax can form bonds with organic compounds that have alcohol groups (OH) in them. Borax links the individual chains by hydrogen bonds. This process is a form of cross linking. When borax is added to water, the compound $B(OH)_4^-$ forms. This compound then reacts with the polyvinyl alcohol molecules. The resulting structure is—

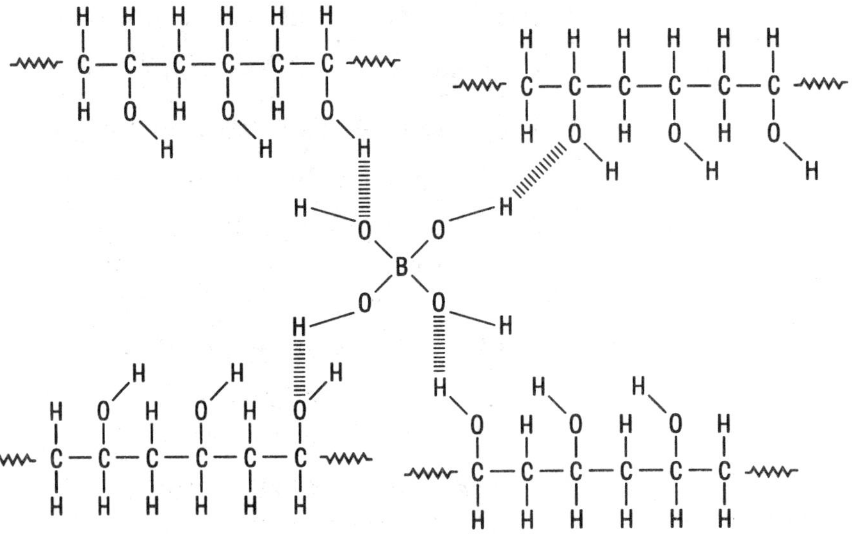

The hydrogen bonds (dashed lines) can occur between an oxygen on the alcohol group and a hydrogen on the $B(OH)_4^-$ group, or they can occur between an oxygen on the $B(OH)_4^-$ group and a hydrogen on the alcohol group. The original polymer chains that were free to move are now bound together by the cross linking. Cross linking restricts the polymer movement and creates a gel. The hydrogen bonds are weak compared to normal bonds and can break and reform as the gel flows.

Safety Precautions

- Before beginning this lab, get a copy of the Material Safety Data Sheet (MSDS) for polyvinyl alcohol. Read this material and observe all cautionary statements.

LAB PROCEDURE

Method

Put on your lab apron and goggles.

1. Observe the solution of polyvinyl alcohol in the stock bottle. Record your observations in your *ABC* notebook. Notice especially how easily it flows.
2. Observe the solution of borax in the stock bottle. Record your observations in your *ABC* notebook. Notice especially how easily it flows.
3. Measure 3 ml of borax solution in a 10-ml graduated cylinder. Add two drops of food coloring to the borax solution. Pour the borax solution into a clear plastic cup.
4. Measure 25 ml of polyvinyl alcohol in a 50-ml graduated cylinder and add to the clear plastic cup.
5. Use the wooden splint to stir the borax and polyvinyl alcohol solution until the mixture becomes very thick. Observe the properties of the mixture. How does the mixture differ from the original solutions? Record your observations in your *ABC* notebook.
6. After the gel forms, lift some of the gel with the wooden splint. Observe the behavior of the gel. Record your observations in your *ABC* notebook.
7. Roll some of the gel into a ball in your hands. Set the ball on the lab bench and observe what happens. Is the gel a solid or a liquid? Record your observations in your *ABC* notebook.
8. Hold some of the gel with your fingers and let it run between them. How long will the gel flow? As the gel flows, touch it to see if there is a temperature change.

Pull on the mass with a sharp jerk. What happens? Record all your observations in your *ABC* notebook.

9. Write on a piece of paper with a water-soluble pen. Press the gel against the writing and peel the gel off. Describe in your *ABC* notebook what happens. Does the ink have to be water soluble?
10. Take some of the gel and put it into another plastic cup. Add table salt. Describe in your *ABC* notebook what happens.

Cleanup Instructions

- The gel can be carried away in the cups in which you made it. It can also be sealed in plastic bags and thrown into the trash. Do not put the gel in your mouth or allow it to get into your hair, on your clothes, or in the carpet.
- Clean any gel from your clothing or desks with warm water, and wipe dry with paper towels.
- Wash your hands with soap and water after handling the gel.
- Wash and dry the glassware and return it to its proper storage location.

Calculation

No calculations are required for this lab.

WRAP-UP

Conclusions

1. Explain what happened to the polyvinyl alcohol solution when you mixed it with the borax solution.
2. Was the gel a solid or a liquid? Give evidence for your answer.

Challenge

3. What happened to the gel when you added table salt to it? Why?

SUBUNIT 5

COMPOSITES

THINK ABOUT IT

- How do the latest models of skis and tennis rackets compare with older, wooden rackets and skis?
- How do the latest racing bicycles compare with older models?
- What properties do the materials that make the equipment shown have in common?
- Are any other vehicles or objects made out of these kinds of materials?

SUBUNIT OBJECTIVES

After you complete this subunit, you will be able to—

1. Research products made from composite materials to find out their structure, components, and properties.
2. Analyze the effect on a composite's performance of using different kinds of structural components.
3. Summarize the advantages and disadvantages of different matrix materials in a composite.
4. Devise a table that relates composite products and the processes used to make them.
5. Compare the properties of a composite material with those of one of its components, using accepted test methods.
6. Demonstrate a method for making fiberglass.
7. Compare the strength of two related materials.

PROCESS SKILLS

You will use these skills in lab—

- Measure length and width using a ruler.
- Measure volume using a graduated cylinder.
- Test performance of a material.

What Is a Composite?

Materials Science Briefing

"Miss Martinez, the Senator is ready for that science briefing now," came the voice over the intercom. Maria grabbed her folder and headed toward the inner office. The Senator was friendly, but looked tired as Maria began giving her mini-lecture on some of the latest developments in materials science. After her summary was over, the Senator sat for a moment, looking over some of the charts Maria had presented.

"Tell me a little more about these composite ***materials,****" requested the Senator. "Why, with all the materials available, are composites necessary? Are they less expensive? This isn't like putting soy flour in hamburger patties, is it?"*

Maria smiled. "No, Senator, it really isn't. Let me explain composites by using the example of a spacecraft. Consider the characteristics, or properties, that are required of a vehicle that might orbit another planet.

"First, it has to take off, then climb through the different levels of Earth's atmosphere and into the cold regions of space. Then it has to withstand a scorching reentry into Earth's atmosphere. It has to be lightweight enough to travel, yet strong enough to carry fuels and instrumentation while hurling through space at incredible speeds. These extreme conditions cannot be withstood by ordinary metals, ceramics, or polymers.

"So, if you put two or more of those together, you can get a material that can do things they can't do by themselves, is that it?" asked the Senator.

"That's right," said Maria. "You might be able to achieve the strength, heat resistance, and corrosion resistance of some ceramics, say, but without the brittleness associated with ceramics."

"Because you combine the ceramic with something else?" guessed the Senator.

"Right again," said Maria. "Senator, have you been reading up on this before I got here?"

"No, Miss Martinez, you just do a good job of setting me up for the right answer. Now, about this bill supporting research in advanced materials. I assume you have a recommendation . . . ?"

A composite material is, as you might guess from the preceding conversation, a combination of two or more distinct materials. Composites are designed to have properties that work better for a given purpose than the properties of their individual components. For example, glass fibers break easily. Polyester is very flexible.

Combining these two materials in a certain way gives a strong and stiff material called fiberglass. The glass fibers stiffen the polymer, and the polymer surrounds the fibers to keep them from breaking. Each component material has distinct properties. When they are combined, the resulting material has properties different from either of the original materials alone.

Composites are different from alloys and copolymers. (Alloys are metallic materials composed of two or more elements, usually metals. Copolymers are composed of two or more different kinds of monomers.) If you slice open a piece of composite material, you can see the different materials of the composite with your unaided eye. With alloys and copolymers, the components are mixed together at the molecular level and can't be distinguished. A composite is like spaghetti and spaghetti sauce, Even when they are mixed together, it's easy to see the individual ingredients, Alloys and copolymers are like chocolate shakes. When you combine the ice cream, milk, and chocolate, you can't distinguish the separate ingredients.

Activity 5-1

- Get a piece of circuit board material, a magnifying glass, and two pairs of pliers from your teacher. Use the pliers to break off corners of the circuit board until all the layers split and reveal the reinforcing fabric.
- Use the magnifying glass to examine the exposed surface. Draw a sketch in your *ABC* notebook of the reinforcing fabric. What direction(s) do you think the reinforcement makes stronger?

Composite Structure

Several different materials can be combined to form composites. These materials can be joined in several different ways. However, most composites have a similar structure, with three parts:

- Body component—called the **matrix** (such as the polyester in fiberglass)
- Structural or reinforcing component—in the form of fibers, flakes, or particles (such as the glass in fiberglass)
- Nonstructural component—often called filler

The Matrix

The matrix gives the composite its basic form. It also binds the structural or reinforcing components together and keeps them in place. Often, single pieces of the reinforcing components would break easily if they were not protected from damage and the environment by the matrix. When the reinforcing components are together in the matrix, they are able to withstand more stress than the matrix material alone can. When a load is applied to the composite, the matrix readily transfers this stress evenly among the reinforcing components (Figure 5-1). Therefore, the composite can withstand a greater load or stress than the unreinforced matrix material.

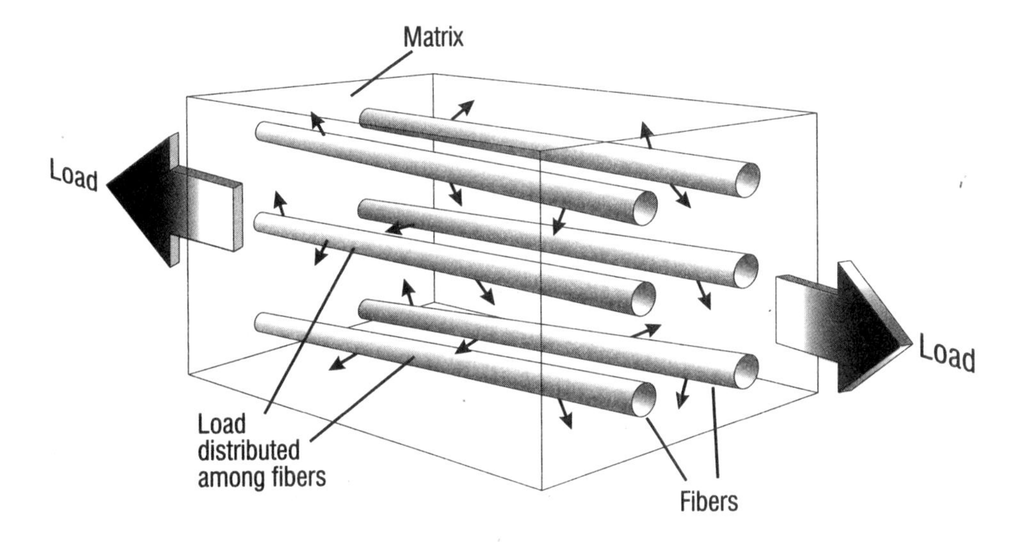

Figure 5-1 Cross section of belted tire under a load

Structural or Reinforcing Component

The structural or reinforcing component is usually in the form of fibers, flakes, or particles. These give the composite its internal structure and properties. When the function of the component is to give the composite its strength, stiffness, or both, the component is called the reinforcement.

Some typical materials used as the reinforcement component are glass fibers or flakes, graphite (carbon) fibers, Kevlar (aramid) fibers, cellulose, metal particles, ceramic particles, and aluminum flakes.

Filler

Filler materials are added to the matrix for reasons other than reinforcement. For example, fillers may be added to improve fire-retarding properties or to increase bulk. When a bulking agent is added to the matrix, less matrix material is required and the cost of the composite is reduced. Filled composites also shrink less than unfilled composites, so the dimensions of a molded part can be controlled more precisely. A filler may also be used to

- Control viscosity to make manufacturing easier
- Produce a smoother surface
- Increase heat resistance
- Increase electrical conductivity
- Increase water resistance

Some typical filler materials include calcium carbonate, clay, mica, glass microsphere (tiny glass beads), talc, and sand.

The basic matrix-reinforcement or matrix-filler structures can be varied. Reinforcements and fillers can be used together to give a desired set of properties. Also, two or more kinds of reinforcing fibers can be combined to meet specific property or product needs. When reinforcements and/or fillers are combined, the composite is called a hybrid.

A good example of a product made with a hybrid composite is a glass graphite epoxy archer's arrow. The arrow shaft is a composite made of glass and graphite fibers in an epoxy plastic matrix. The graphite fibers' stiffness and the glass fibers' strength allow the shaft to be made with a smaller diameter than an equivalent aluminum shaft. Having a smaller-diameter shaft reduces drag on the arrow in flight. Also, the stiffness of the graphite helps reduce the vibration at release and during flight. The result is an arrow with greatly improved performance.

Activity 5-2

- Visit one or more stores that sell sporting equipment, camping equipment, bicycles, boats, and so on. (You may want to divide the class and let each small group visit a different store.) Examine the recreational equipment that appears to be made of a composite material. These are often ultralight as well as strong.
- Identify the companies that make this material (from the product tags or salesperson). Most retail stores will have the company names and addresses on file.
- Write or call the producer to obtain product information. Find out, if possible, the component materials, structure, properties, and the process used to make it.
- As a class, begin work on a table of information about these products. At this time, fill in three columns: product name, component materials, and structure. Save this table of information for use in the Activity 5-6 and the Unit Wrap-Up Activity.

What Are the Types of Composites?

Composite materials are often classified based on the form of structural component or the type of matrix material.

Form of Structural or Reinforcing Component

Structural or reinforcing components may be in one of these forms:

- Flakes
- Particles
- Fibers
- Laminar materials

Flake Composites

A flake composite is made up of flat flakes, similar to those shown in Figure 5-2.

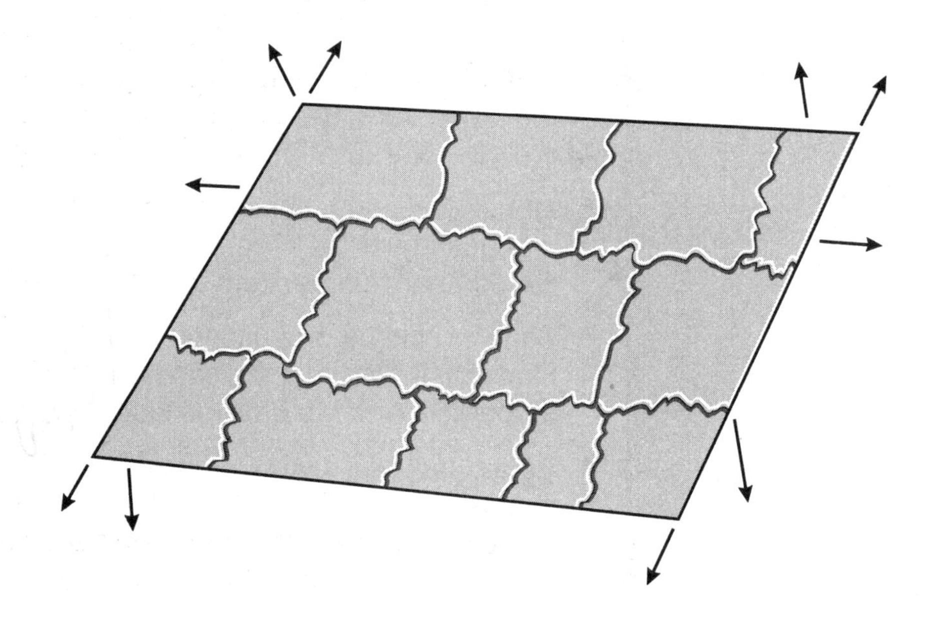

Figure 5-2 Flake composite structure

The flakes are nearly parallel and overlapping, like the shingles on a roof. The flakes give what is called planar **isotropic** (ī′sə-trō′pĭk) reinforcement. That is, they reinforce equally in all directions in the plane of the flakes. However, they provide little or no perpendicular reinforcement. In addition to reinforcement, the overlapping structure of the flakes provides other important properties. The overlapping flakes form a very effective barrier against moisture. Metal flakes that not only overlap but actually touch can be used to make the composite electrically conductive. Nonconductive flakes, such as glass, can provide electrical or heat insulation.

Flake composites are the least used structural form. Manufacturers have difficulty making them and therefore look for alternatives. One problem is obtaining correct alignment of the flakes. Another problem is making the flakes the desired shape and size without weakening flaws. Also, the number of flake materials is limited.

The most commonly used flake materials are aluminum and glass. Aluminum is used mostly for corrosion resistance and heat dissipation. Electronics manufacturers use glass flakes in printed circuits and molded insulators.

Particle Composites

A particle composite consists of hard particles embedded in a softer matrix. Figure 5-3 shows a particle reinforcement.

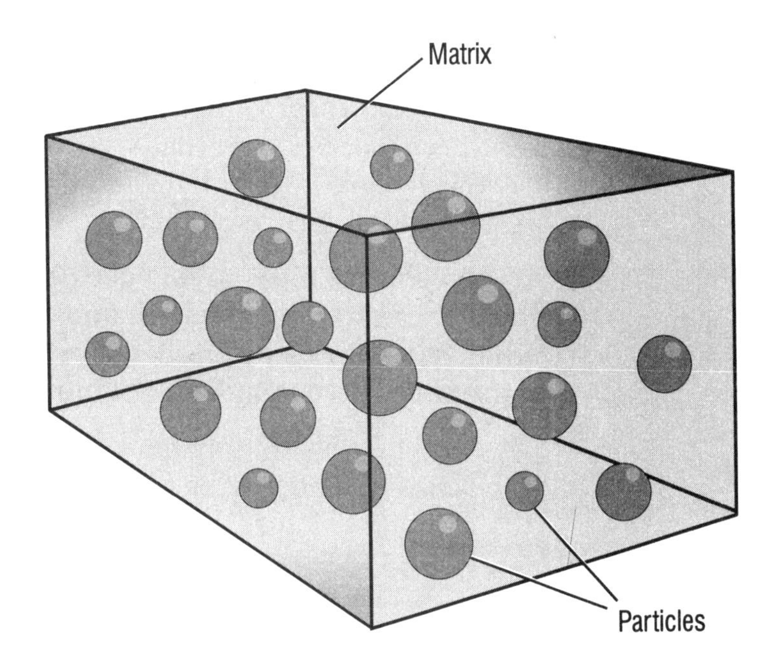

Figure 5-3 Composite particle reinforcement structure

The particles provide approximately the same reinforcement in all directions. This reinforcement is called isotropic. Notice that the particles are different sizes. Not every particle composite has different-sized particles—some are made with a uniform particle size. However, different-sized particles can be an important design factor. The smaller-sized particles fit in between the medium ones, which in turn fit between the larger particles. The result is a compact structure with more reinforcement and less matrix material. A common example is concrete. The sand and gravel particles reinforce the cement matrix material.

Fiber Composites

Fiber composites consist of fiber reinforcements in a matrix material. These composites have properties that make them among the most versatile structural forms. The properties of fiber composites can be tailored to meet a wide variety of needs depending on these factors:

- Fiber (and matrix) materials used
- Fiber length

- Fiber orientation (the direction in which fibers are set)
- Fiber-matrix bonding
- Fiber-matrix ratio

Two kinds of fibers may be used in a composite—natural and/or synthetic. Two commonly used natural fibers are asbestos and cotton. Builders use asbestos composites primarily for fireproofing and insulating. (The use of asbestos is decreasing because airborne asbestos particles are known carcinogens.) Cotton fibers provide strength, weather resistance, machinability, and toughness.

Glass is the most widely used synthetic fiber. Glass fibers offer excellent strength and electrical properties. Other types of synthetic fibers are the advanced fibers such as graphite, boron, and the polymers known as aramids.

- Graphite is a highly pure and ordered form of carbon whose fibers provide high strength and stiffness. The crystal structure of graphite is shown in Figure 5-4.

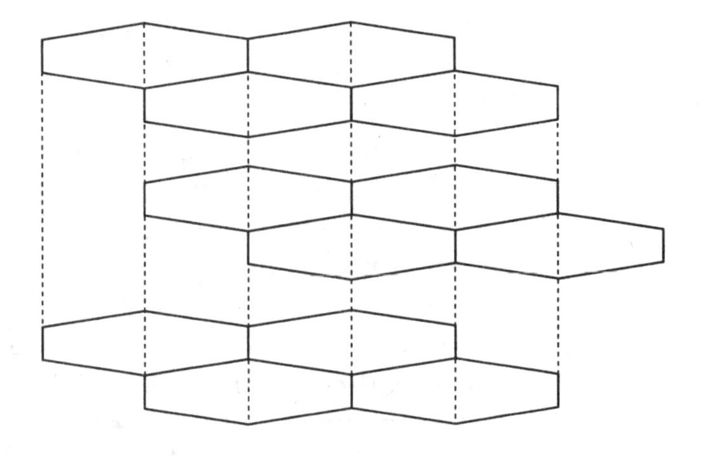

Figure 5-4 Crystal structure of graphite

- Boron is not used by itself, but is often deposited on an underlying filament of tungsten or carbon. Silicon carbide (SiC) is sometimes coated on the outside of the boron. Boron fibers are often used in aluminum matrices and, if not coated, they would quickly degrade upon contact with the molten aluminum.
- Aramids are lightweight, strong, heat-resistant synthetic polymers. DuPont's Kevlar 29 and 49 are among the aramid fibers commercially available. The chemical structure of Kevlar fibers is shown in Figure 5-5.

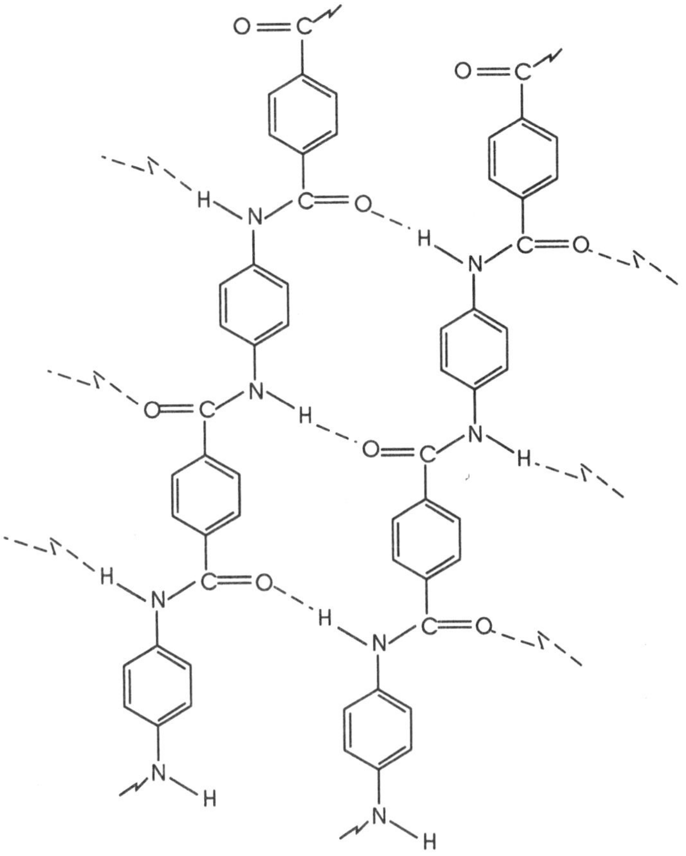

Figure 5-5 Chemical structure of Kevlar

Other often-used fibers include ceramics such as silicon carbide and aluminum oxide (Al_2O_3). Silicon carbide maintains its tensile strength above 650°C (1200°F). Aluminum oxide has excellent strength up to about 1370°C (2500°F). Both fibers can reinforce metal matrix composites in which carbon or boron fibers react with the metal. These are a few of the fiber materials available for fiber composite use.

Fiber Properties

The many different fibers available can add a variety of properties to composites. Fibers may have good chemical resistance, electrical insulation, or high-temperature performance. However, in most fiber composites, the fibers function mainly as reinforcements. Therefore, their most important properties are strength and stiffness, The ability of a fiber to carry load without breaking is called strength. The ability of a fiber to resist bending or deflection is called stiffness (or modulus).

The most commonly measured properties of fibers are their tensile, compressive, and shear strengths and stiffnesses. Tensile loads tend to pull the ends of a material away from each other. A stretched rubber band is under tensile loading. Compressive loads push two ends of a material together. Crushing a soft drink can is an example of a compressive load. Shear loads break a material by pushing its surfaces in opposite directions. Turning an Oreo cookie to separate the two halves puts the frosting inside under a shear load. Table 5-1 shows typical tensile strengths and moduli (stiffnesses) for the fiber materials discussed earlier.

Table 5-1. Fibers and Their Properties

Fiber	Tensile strength	Tensile modulus
Aramid		
Kevlar 29	5.5×10^5 psi	9×10^6 psi
Kevlar 49	5.5×10^5 psi	1.9×10^7 psi
Boron		
Tungsten core 4 mil thick	5.0×10^5 psi	$5.8\text{-}6.0 \times 10^7$ psi
Carbon core 4.2 mil thick	4.75×10^5 psi	5.3×10^7 psi
Glass		
E glass	5.0×10^5 psi	1.05×10^7 psi
S glass	6.5×10^5 psi	1.24×10^7 psi
Graphite		
Low-cost High-strength (LHS)	1.75×10^6 psi	$3.0\text{-}3.4 \times 10^7$ psi
Intermediate-modulus (IM)	1.71×10^6 psi	$3.3\text{-}4.0 \times 10^7$ psi
High-modulus (HM)	1.85×10^6 psi	$5.0\text{-}5.8 \times 10^7$ psi
Silicon carbide (tungsten or carbon core)	5.5×10^5 psi	$5.0\text{-}6.0 \times 10^7$ psi
Aluminum oxide	2.0×10^5 psi	5.5×10^7 psi

Activity 5-3

- Use Table 5-1 as a source of information to develop a graph of tensile strength and tensile modulus for all of the materials listed.
- Using both the table and the graph, answer the following questions as a class:
  - Which material appears to have the greatest tensile strength? The least?
  - Which material appears to have the greatest stiffness? The least?
  - Is there an apparent relationship between tensile strength and tensile modulus? If so, what is it?

Engineers must also consider the cost of the fiber when they design a composite. Generally, the higher the strength and stiffness of the fiber the higher the cost. Microscopic flaws in a fiber decrease its strength and stiffness, and fabricating an advanced fiber with minimal flaws is difficult.

Fiber Orientation

The concepts of isotropic reinforcement and planar isotropic reinforcement were discussed earlier. Fiber composites increase the possibilities of reinforcement. The fibers provide reinforcement almost entirely in the direction of their length. Therefore, the orientation of the fibers greatly affects the properties of the composite. Figure 5-6 shows some possible reinforcement patterns.

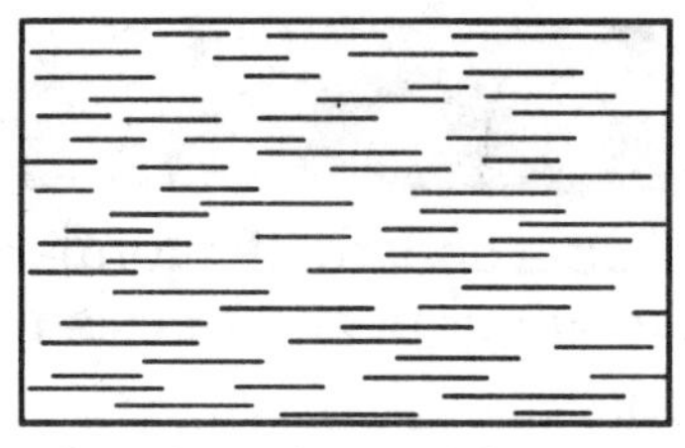

a. One-dimensional reinforcement

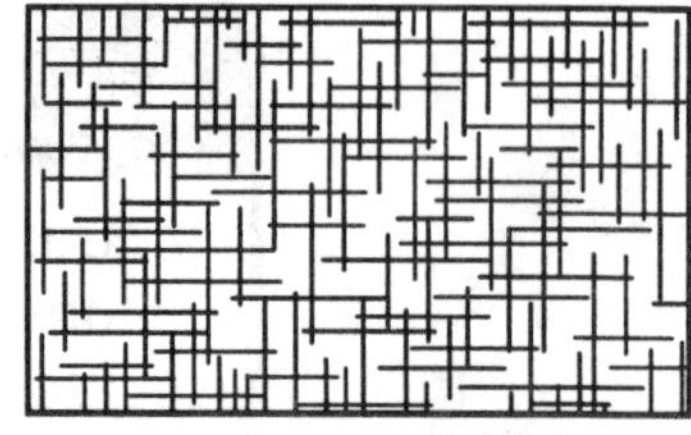

b. Planar reinforcement

c. Planar isotropic reinforcement

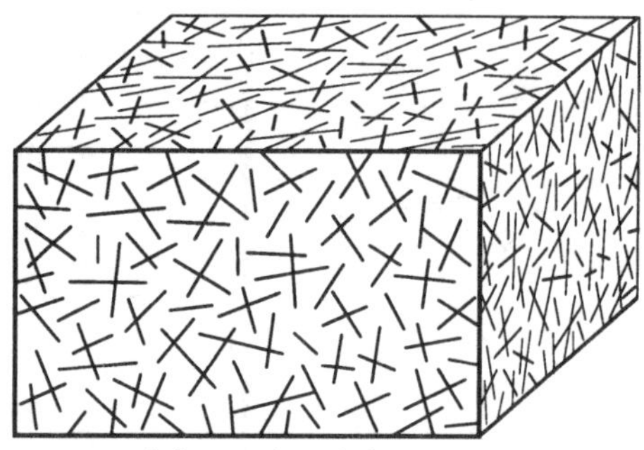

d. Isotropic reinforcement

Figure 5-6 Fiber reinforcement patterns

In Figure 5-6a, the fibers are all oriented in the same direction. This type of reinforcement gives one-dimensional (unidirectional) reinforcement. Figure 5-6b shows fibers oriented both lengthwise and widthwise in the same plane. Equal strength in both directions results and is called two-dimensional, or planar reinforcement. The next picture shows fibers oriented randomly in the plane. This type of reinforcement is called planar isotropic, just as when flake reinforcements lie parallel in the same plane. Finally, fibers placed randomly in all directions throughout the matrix result in isotropic reinforcement, much the same as in particle composites.

The fibers in the matrix act as reinforcing cables, so an important fact to keep in mind when designing a composite is that the more random the reinforcement, the less strength there is in any one direction. A unidirectional composite has almost all of its strength in one direction and has very little strength in any other direction. An isotropic composite is equally reinforced in all directions. However, the strength of an isotropic composite in any one direction is much less than the strength of a uni-directional composite in a single direction.

Fiber Length

Fiber reinforcements can be either continuous or discontinuous. Continuous fibers run from one edge of the composite to the other with no break in the fiber. Discontinuous fibers are normally one inch or less in length and are separate pieces in the composite. Continuous fibers are easier to orient in a given direction; however, discontinuous fibers oriented carefully give greater strengths than continuous fibers. The reason for this is that shorter fibers have fewer weakening flaws. Very short fibers, called whiskers, come closest to the maximum possible strength of the materials from which the whiskers are made.

Fiber Shape and Form

Fibers with circular cross sections are the most common type. However, other types of cross sections do occur. These types are rectangular, hexagonal, polygonal (many sided), irregular, and annular (hollow cylinder).

Fibers come in several forms for use in manufacturing composites. The basic form of continuous fibers is a strand, or bundle of fibers. A roving is a group of untwisted fibers wound on a cylinder, similar to a spool of thread. Woven roving is a drapable fabric that consists of continuous rovings woven in two perpendicular directions. A woven roving mat looks similar to a woven placemat. Chopped strands and a binder material combine to form a chopped-strand mat. Felt could be considered a type of chopped-strand mat. All of these forms of fibers combine with a matrix material to form a composite.

Fiber-Matrix Bonding

The interaction between the matrix and fiber is what makes a fiber composite work. Under a load, the weak matrix material deforms and distributes the stress among the strong fibers. The interaction takes place only if the matrix and fibers are tightly bonded together. Thus, the fiber-matrix bond is extremely important. (In this instance, the word bond does not imply a chemical bond.) If this bond is weak, the matrix will slip past the fibers and the load will fall on a few fibers. When these fibers break, the load will be passed to a few more fibers, until all the fibers break. The effect would be the same as if it had no matrix.

The fiber-matrix bond depends on the types of materials used for the matrix and the fibers. Sometimes the matrix can be bonded directly to the fibers. Other times, the matrix and fiber materials will not bond strongly to each other. In these cases, another material called a bonding agent is used between the two. The bonding agent is deposited on the fiber so the matrix will adhere more effectively to the fiber. Figure 5-7 shows the fiber, agent, and matrix.

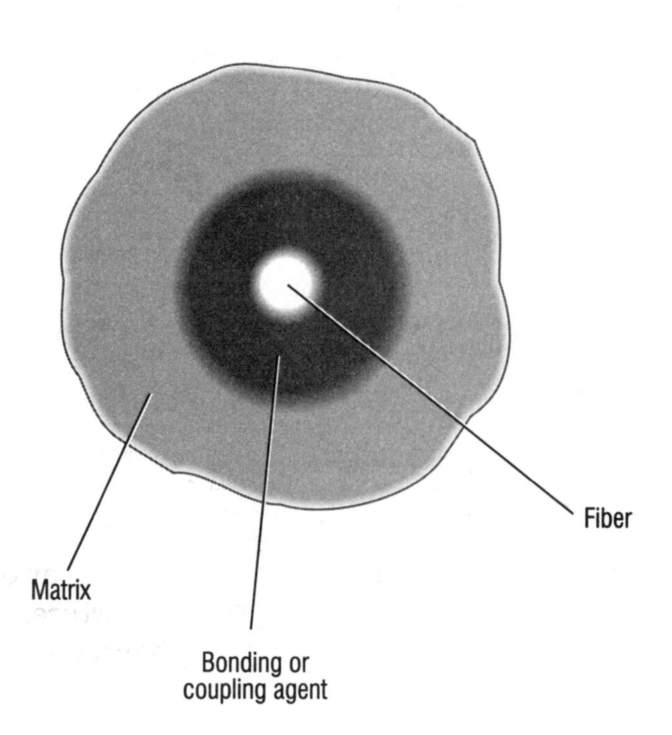

Figure 5-7 Fiber, bonding agent, and matrix of composite

Fiber-Matrix Ratio

Since the fibers contribute the basic strength to the composite, one would assume that increasing the fiber content will improve the properties of the resulting composite. This is generally true—within limits. For example, above a certain fiber content the fiber-matrix bond becomes more important than the percentage of fiber content. If not enough matrix is available to coat the fibers, a weak composite results. Also, short fibers have a greater thickening effect on the matrix than continuous fibers. A fiber-matrix that is too thick cannot be molded or shaped properly. For this reason, continuous fibers are used at higher concentrations than discontinuous ones.

Laminar Composites

A laminar composite is a product that contains two or more materials bonded together in layers. The materials may be similar or quite different. Laminar composites are unusual for two reasons. The first is that they are not based on a matrix-reinforcement type of structure. The layers are simply bonded together. Each layer contributes its own properties to the composite.

Plywood is one of the best-known laminar composites. Snow skis are also laminar composites. They have a wood core, fiberglass and rubber layers, and steel edges.

In some laminar composites, one of the materials completely surrounds the other. The enveloping materials give the products some special property such as corrosion resistance. A typical soup can is tin-plated steel. The tin prevents steel corrosion in the presence of acidic foods (tomato soup). The steel provides strength and stiffness to the can. Some laminar composites combine different types of composites within their structure. A snow ski is an example of this. The fiberglass layers are a composite material, but they also are combined with other layers of materials to form a whole ski.

Figure 5-8 shows a laminar composite that incorporates a honeycomb sandwich structure. Such a structure might form one of the layers in a helicopter blade or other aircraft body parts.

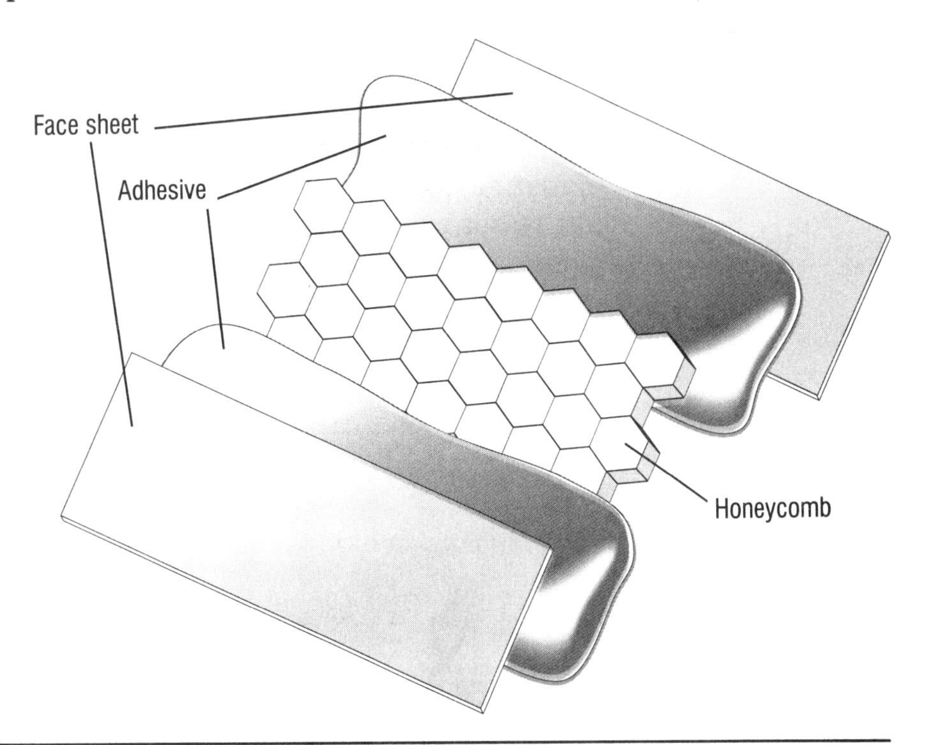

Figure 5-8 Honeycomb sandwich structure

The face sheets are made from graphite, fiberglass or aramid composites, or aluminum. The honeycomb core is usually made from aluminum, titanium, or a paperlike material called Nomex. As you can see, a honeycomb structure is incorporated in many of the advanced materials. Honeycomb structures are used when high strength is needed without a lot of added weight.

Type of Matrix Material

The matrix of a composite has three main jobs to do. It transfers stresses from the matrix to the reinforcement. It protects the reinforcement from damage. It provides a barrier against adverse environments. Although the reinforcement carries the majority of the tensile load, selection of a proper matrix is still important. The matrix provides support against fiber buckling under compressive loading, and therefore contributes to the compressive strength of the composite. The interaction between the matrix and the reinforcement is also important. If the interaction is noneffective, the load will fall entirely on the reinforcement. Last, the processability of a composite depends on the physical characteristics of the matrix, such as viscosity and melting point.

Metals, ceramics, and polymers can be used as composite matrix materials. Polymers are by far the most common matrix material. However, with the demand for high-temperature applications, research and development for ceramic- and metal-matrix composites are increasing.

Aluminum is the most common matrix material in metal matrix composites. Titanium and magnesium are also used to a lesser extent. Ceramic materials such as borosilicate glass are also used.

Activity 5-4

- As a class, develop lists of the possible advantages and disadvantages of using metal as a matrix material. Do the same for ceramic materials. (Base your list on what you have learned so far [in Subunits 2 and 3] about metals and ceramics.)
- Compare your lists with those provided by your teacher. Discuss the differences.
- Record your findings in your *ABC* notebook for use in the Unit Wrap-Up Activity.

Polymer matrix materials are divided into two categories: thermoplastics and thermosets. Remember from the polymer subunit that a thermoplastic has individual linear molecules with no chemical linking or bonding between them, much like a bowl of spaghetti. The molecules are held together by weak bonds that can be easily broken under heat and pressure. Upon cooling, the molecules harden in their new positions and restore the weak bonds. Thus, a thermoplastic polymer can be softened, melted, and reshaped many times.

The biggest advantages thermoplastics have over thermosets are high impact strength and fracture resistance. These in turn give the composite great damage resistance. Other advantages of thermoplastics are

- Unlimited shelf life at room temperature
- Shorter fabrication time
- Postformability (polymer can be reformed after initial hardening)
- Ease of handling (no tackiness)

Even with these advantages, the development of thermoplastic matrix composites has been slower than thermosets. Thermoplastics have higher viscosities, which makes incorporation of reinforcements difficult. The polymer does not wet the surface of the fiber as easily, which may result in a poor fiber-matrix bond. Thermoplastic materials have lower thermal stabilities than thermosets, and therefore are not used in high-temperature applications.

Thermoplastics also have poor solvent resistance compared to that of thermosets.

Thermoset polymers, on the other hand, have molecules that are chemically joined by cross links. These linkages form a rigid, three-dimensional network structure. When these cross links form, the thermoset polymer cannot be melted and reshaped by applying heat and pressure. Because of this, thermosets tend to be more brittle than thermoplastics. Thermosets are generally added to reinforcements in the liquid state (resin). A curing agent is added to help speed curing (hardening) of the resin. Many thermosets have long curing times so the polymer goes through a tacky stage before hardening completely.

Care must be taken when using curing agents, as **hypersensitivity** can result. Hypersensitivity causes an allergic skin reaction for some people. Other people become so sensitive to the curing agent that they cannot be in the same room without suffering adverse symptoms.

Because thermoset polymers do not soften, they have traditionally been used as the matrix materials for polymer composites. Starting materials are usually low-molecular-weight materials with low viscosities. Since the viscosity of the polymer is low, a good fiber-matrix bond can be achieved without the need for heat or pressure. Thermoset polymers have good thermal stability and chemical resistance. Two disadvantages of thermosets are their limited storage life at room temperature (after mixing the resin and curing agent), and their long fabrication time in the mold (where the reaction is carried out).

The characteristics of a polymer matrix composite are summarized in Table 5-2.

A special class of polymers is the elastomers (or rubbers). These materials have high flexibility and can also be used as matrices. A well-known example is the rubber in an automobile tire. Possible reinforcements include steel, nylon, or aramid fibers (belts). These tires combine the flexibility and shock absorbance of the rubber with the strength of the fibers. Reinforced garden hoses and fan belts are other examples.

Table 5-2. Characteristics of Polymer Matrix Composites

Pros	Cons
Good in-plane strength and stiffness	Poor high-temperature performance
Low density	Moisture sensitivity
Ease of fabrication	Low through-thickness strength and stiffness
Relatively low cost	Poor impact resistance (thermosets)
Corrosion resistance	Special processing equipment required
Fatigue resistance	
Low coefficient of thermal expansion	
Can be made electrically conductive or nonconductive	
Can be made thermally conductive or nonconductive	
Mature technology	
Much in-service experience	

Manufacturing of Composites

An important part of the application of a material is a cost-effective and reliable method to manufacture it. One benefit of composites is the ability to create a "custom design" by proper selection of reinforcement material, reinforcement orientation, and matrix material. But if no practical way to manufacture the composite exists, the full design benefit may not be realized.

Because many forming methods have been discussed in the other subunits, the techniques used are summarized here in Table 5-3.

Table 5-3. Manufacturing Methods for Composites

Matrix	Manufacturing Method
Metal	Ingot metallurgy (foil/fiber/foil) Powder metallurgy Casting
Ceramic	Cold pressing and sintering Devitrification Slip-casting
Polymer	Hand lay-up Bag molding Compression molding Sheet molding Bulk molding Wet system compression molding Reinforced thermoplastic sheet compression molding Pultrusion Filament winding Prepegs

Activity 5-5

- Working in groups of two, let each group select one of the manufacturing methods listed in Table 5-3.
- Using a teacher-provided handout and/or library or World Wide Web resources, develop a presentation on the selected method. Work with your partner in developing and giving the presentation. Use diagrams, physical models, verbal descriptions, or simulations to show how the process is carried out.
- With your partner, develop one or two test questions on the process that you are teaching to the class. (Your teacher will assemble the test questions into one test to be taken by the entire class.)
- Take the test on manufacturing methods.
- Score the test and determine which methods were best understood (based on the test items that were missed least).

Activity 5-6

Using the table of information you developed in Activity 5-2, try to determine what method of manufacturing was used for the products listed. (Some may be obvious; others may require checking with the manufacturer.)

Testing Composites

CAREER PROFILE: COMPOSITES TESTING ANALYST

Lu C. is a testing analyst for an aerospace company. She tests metals and composite materials used to manufacture aircraft.

"One of the main instruments I use is the scanning electron microscope," says Lu. "I use it to examine surface conditions in composite materials."

Lu explains that composite materials are combinations of two or more materials used together. "One composite material that everybody knows is plywood," she begins. "In plywood, thin layers of wood are laminated together with the grain of the layers going in different directions. The result is a sheet of wood that is stronger than a piece of wood with the grain going in one direction only.

"Graphite epoxy is the composite material that I test most frequently. Unlike plywood, which is a layered composite, graphite epoxy is a particle composite. Particles of graphite are distributed throughout an epoxy matrix. A matrix is a kind of open framework. The epoxy gives the material a degree of flexibility, and the graphite particles give it greater strength. That's a simplified description, but it gives you the general idea."

What is Lu looking for when she examines composite materials using the electron microscope? "Primarily I'm looking for surface defects—areas where the surface conditions are irregular and the material might be vulnerable to stress as a result.

"Another aspect of my job is to analyze material

failures such as fractured parts. I use a technique called fractography to do that. Occasionally, I am called upon to inspect air-crash debris. Also, sometimes I do what is called elemental analysis of materials," Lu continues. "I run tests to determine the elements contained in a given sample."

Lu is not free to describe some aspects of her job because the work is being carried out for the national defense and is classified information. Lu has an associate degree in nondestructive testing from a community college program.

Testing a material is important to make sure that it meets the needs of the application. The more critical the use of the material, the more testing is required. Also, a new material in a new application requires more testing than a well-known material in an established application.

There are two basic categories of testing: destructive and nondestructive. In destructive testing, the test material is destroyed in the process of testing. In nondestructive testing, the test material is not damaged by the testing process.

Destructive Testing

Destructive testing done in a research lab helps scientists to understand how a new material will respond to different conditions. Destructive testing is also used as a quality control measure for final products in a company. A small number of units selected from the finished products are tested to the point of destruction. Sometimes specimens made specifically for testing are produced along with the product. These specimens are called coupons. Coupons can be made in two ways. The coupon can be machined directly from a nonessential area of an actual finished product. Or the coupon can be made from the same materials used for a finished unit. Coupons are made under exactly the same conditions as the production units.

The destructive testing techniques used to evaluate composites are basically the same as the techniques used for metals, ceramics, and polymers. The tension test is the simplest and most common of destructive test. A load is applied through the end of a specimen. (That is, the specimen is pulled.) This test can determine tensile strength and elastic modulus (stiffness). The specimen shape has a large effect on test results. Figure 5-9 illustrates three common test shapes. Notice that the ends of all three are thickened somewhat. The thickening reduces the chance of failure at the end areas.

Compressive testing is another destructive test. Compressive tests are very similar to tensile tests, but the sample is subjected to compression instead of tension. Thin, flat samples like the ones used in tensile testing often buckle under compression, which interferes with test results. Therefore, blocks and bars are often used as specimen shapes.

Another problem with compression testing is premature failure at the ends of the sample. One type of end failure that occurs with both metal and polymer matrices is called brooming. This happens when the load is in the fiber direction. The matrix is crushed and the fibers spread out. The result is a sample with an end that looks like a broom.

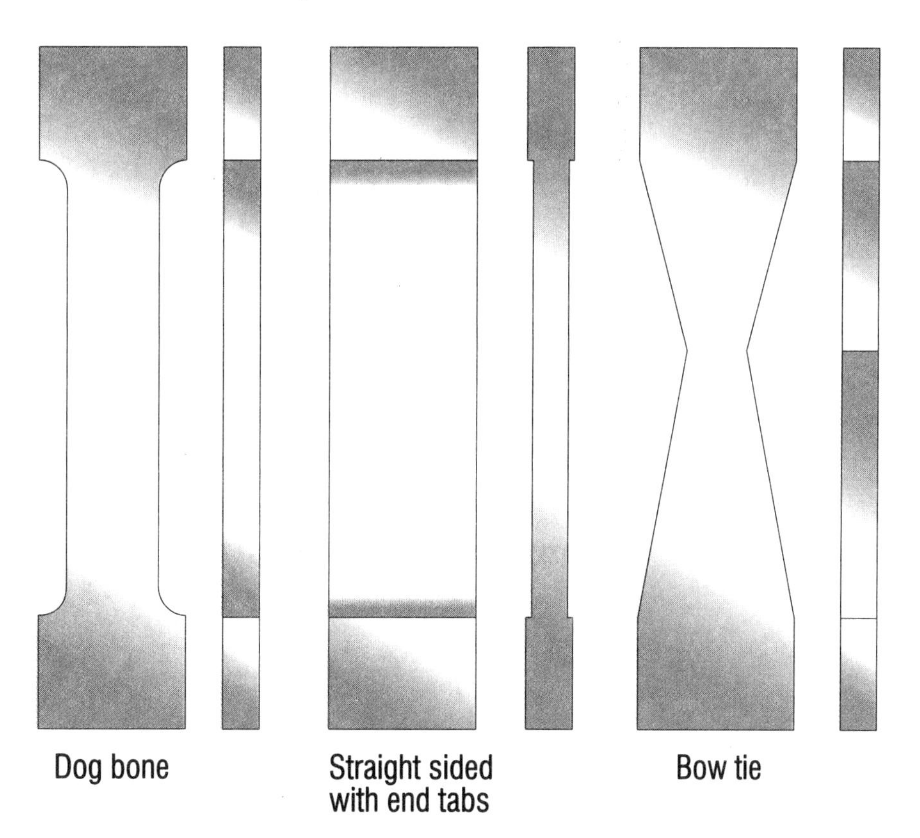

Figure 5-9
Test shapes

Flexural tests measure bending properties. Three-point and four-point tests are the common forms of flexural testing. Figure 5-10 shows the two types.

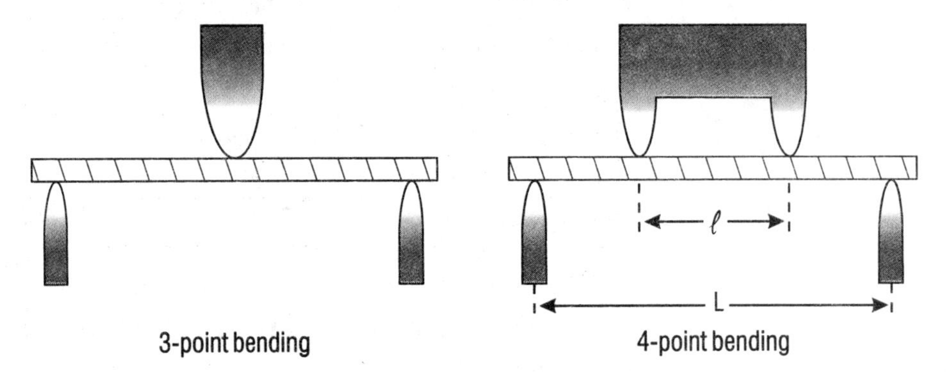

Figure 5-10
Types of flex testing

In three-point testing, interlaminar (between layers) shear stress occurs and is uniformly distributed along the length of the specimen. The interlaminar shear strength of a composite is relatively low, so there is a high potential for shear stress to destroy the specimen before its bending properties fail.

In four-point loading, two symmetric loads are placed on the sample. No shear stress occurs between the loads—it occurs only between the load points and the nearest support. Between the loads, the sample fails in flexure, not interlaminar shear.

When a part fails, a technique called fractography is used to determine how and why the failure occurred. Fractography is a microscopic analysis of the fracture surface. It identifies the crack origin and crack growth direction. Fractography can also be used to determine the type of loading that caused the crack to start.

Nondestructive Testing

Nondestructive testing (NDT) is a growing field. Many NDT techniques are used to ensure quality control. In some cases, NDT represents the only practical way to test a part from a cost standpoint; very expensive parts cannot be destructively tested.

Table 5-4 shows some common composite defects and NDT techniques used to detect them.

Table 5-4. Composite Defects and NDT Techniques

Defect	Sonic	Ultrasonic	X ray	Microwave
Debonding	X	X	X	
Delamination	X	X	X	X
Undercure				X
Fiber misalignment			X	
Damaged fibers			X	
Variation in resin				
Thickness	X	X	X	
Density		X	X	
Voids	X	X	X	X
Porosity	X	X	X	X
Fracture	X	X	X	X

Sonic testing uses a metal disk called a "coin" to detect delamination (separation of a composite layer from an adjoining layer) and large voids (air entrapped in the resin). Tapping the coin along the part surfaces produces sounds. A well-bonded part produces a sharp, ringing sound. Delamination or voids produce hollow or dull sounds. The reliability of this technique depends heavily on the technician's experience.

Ultrasonic testing uses sound waves at frequencies (100 kHz to 25 mHz) that cannot be heard by the human ear. A source, such as a pulse generator, transmits electrical impulses to a transducer. The transducer converts the impulses into vibrations. The vibrations pass into the sample. The ultrasonic wave travels through the specimen, and the reflected energy is detected by another transducer. The energy is converted back into electrical energy and is used to create a visual display on a cathode ray tube (CRT). The visual display is read by a technician.

X-ray inspection is similar to the technique used in medical and dental diagnostics procedures. X rays can penetrate many solid materials. The amount of penetration depends on the thickness, density, and composition of the material. The intensity of the **radiation** that emerges reveals the internal structure of the material.

Microwave techniques are similar to ultrasonic techniques. A source directs microwave radiation at a sample. A detector picks up the energy transmitted through or reflected from the sample. The detector converts the energy into electrical energy to create a visual display. Because of the high frequencies of microwaves (0.5 GHz to 1000 GHz), smaller defects can be detected than with ultrasonics. The higher frequencies allow for better focusing of the radiation on smaller areas.

From Implants to Aerospace—the Uses of Composites

The field of composites is growing rapidly, and the applications are too many to cover in this subunit. Here, we will focus on two areas in which composite materials are being used—medicine and aerospace.

Rebuilt Tendons and Artificial Limbs—Using Composites in Medicines

Living tissues resemble synthetic composites, especially those that are naturally reinforced with fibers. In animals, such fibers are usually made of protein. One example is the connective tissue that makes up a tendon (Figure 5-11). A tendon is a type of connective tissue that links a muscle to a bone. As the muscle contracts, it pulls on the tendon, which draws the bone toward the muscle.

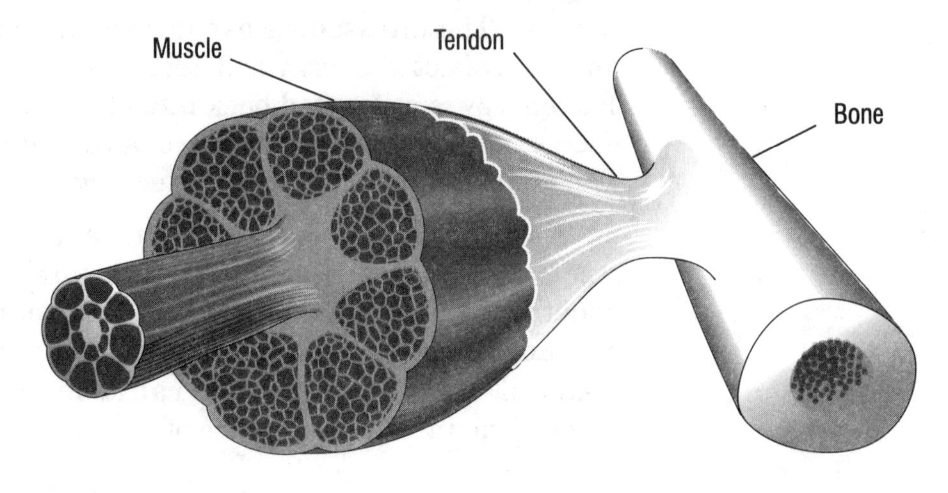

Figure 5-11
Tendon

Tendons are sometimes torn, especially from the stress of physical activity. In a condition known as tendinitis, a tendon has become inflamed and swollen, causing a lot of pain. A tendon that is severely damaged may sometimes fail to heal, even after lengthy treatment. In such cases, an orthopedic surgeon may recommend a special implant made of a polymer matrix and synthetic (carbon or polyester) fibers. The surgeon weaves the implant into the patient's damaged tendon. Soon after surgery, tissues surrounding the implant begin to break down and absorb the matrix material. As the matrix is absorbed, connective tissue regrows around the remaining synthetic fibers.

Another use of new composites in medicine is artificial limbs that will seem more natural to the wearer. To work comfortably, an artificial limb must not only support body weight but distribute it to all parts of the mechanical limb. It must also allow for motion that resembles the way in which bones, muscles, and tendons normally behave during activity. A new model of artificial leg is custom-made from a thermoplastic reinforced with fiber. The new composite, called Fiberod, is molded to match the walking style of the wearer.

Supersonic Transport and Space Shuttles—Using Composites in the Aerospace Industry

Some composite materials offer at least two main properties that are in demand in the aerospace industry: They are lightweight and strong. In making parts for supersonic aircraft designed to travel within the Earth's atmosphere, a primary goal is to reduce weight. The greater the weight of an aircraft, the more fuel is required to move it from one place to another. Therefore, saving weight is saving money. Many of the new composite materials are stronger than steel but lighter than aluminum, and are expected to make up most of the primary structure of the aircraft frame in future designs.

Another property that is important for supersonic transport planes is heat resistance. Because of the high speeds of these aircraft, the temperatures on their surface can rise to 1000°F.

The space shuttle has placed special demands on materials scientists and engineers, because it is repeatedly exposed to high temperatures and high speeds. The nose cone of the shuttle is made of a laminated material often called RCC by engineers. RCC stands for reusable carbon-carbon. RCC consists of layers of graphite-fiber cloth with a carbon matrix or binder and a silicon carbide coating. Each of these three materials in the laminate has a function. The fibers and the binder provide strength and resist high temperatures. The binder provides rigidity. The coating resists oxidation.

Looking Back

Composite materials are combinations of two or more materials. Composites are designed to create a material with properties that work better for a specific use than their components alone would work.

Composites generally are made from a matrix material that gives the product its form and binds the other materials in place; a structural or reinforcing material that is held within the matrix; and sometimes a nonstructural filler material, also held in the matrix.

Structural components may come in the form of flakes, particles, fibers, and laminar materials. The way these materials are dispersed or arranged affects the properties of the composite. Matrix materials may be metal, ceramic, or polymer materials, but polymers are the most commonly used.

Composite materials are usually designed for very specific uses. Often they are required to have extraordinary properties that cannot be found in traditional or advanced materials, ceramics, or polymers. The testing of composites to make sure that they have the intended properties is an important step in their manufacture.

Further Discussion

- What building materials do you know of that could be classified as composites?
- For composite materials to function properly, there must be adhesion between the matrix material and the

structural or reinforcing component. Sometimes the adhesion is very strong; sometimes it is not so strong. What do you think would be the effect of different degrees of adhesion in a composite?

Activities by Occupational Area

General

A Concrete Idea

- Arrange to visit a construction site at the time of a concrete pour. After the major work of the pour is concluded, interview the superintendent or foreman to find out what's involved in managing a successful pour. Find out how the concrete is ordered, how the aggregate size is determined, and what factors contribute to the strength of the concrete after it dries.

Agriscience

Plant Matter Composites

- Investigate the building materials used for housing in some less-developed countries (LDCs). Find out if any of these materials includes plant matter as an ingredient of a composite material (such as straw-reinforced bricks). If plant matter is used, find out what kind and whether crops are grown specifically for the purpose of providing materials for shelter.
- Find out how these materials compare with materials conventionally used for housing in the United States in terms of strength, weather resistance, thermal properties, and durability. A good place to start your research is with the organization Habitat for Humanity.

Health Occupations

Fiberglass Hazards

- Contact the National Institute for Occupational Safety and Health to find out what hazards to workers, if any, are presented by working with fiberglass. Ask questions about workers in fiberglass plants as well as workers who repair fiberglass products (such as people who work in automobile body shops).

Family and Consumer Science

Counter Intelligence

- Visit a building supply company to find out what kinds of laminates are used for kitchen countertops and other work surfaces in the home. Find out if there are significant differences among the different types in the wear resistance of the surface, the strength of the countertop under load, and other properties.

Industrial Technology

Lightweight Bridges

- Contact the Federal Highway Administration and ask about the use of composite materials in bridges. Find out what plans are underway to use composite materials in bridges, where the first such bridges will be built, what materials will be used in the composites, and how the weight and performance of these bridges will differ from those of steel-and-concrete bridges.

LAB 10

MAKING AND TESTING FIBERGLASS

PREVIEW

Introduction

Tommy G. is a materials technician. He works for a chemical company that makes many of the resins used in fiberglass products. Tommy's job is to build and test samples of a fiberglass composite. The performance of the samples during testing provides needed information about the quality of the resins that the company will sell to fiberglass industries.

Tommy builds the sample "layups," as they are called, by manually combining layers of resin-wetted fiberglass. He begins by mixing the resins according to manufacturer's instructions. He then tests each layer of glass fiber with the resin and lays it in place. He uses a roller to roll out air pockets in the layers. The sample is then allowed to "cure" or set up at room temperature.

The second part of Tommy's job is to test the fiberglass layups. He conducts two physical tests—a burn test and an acoustical test.

The burn test involves weighing the sample then burning it to vaporize the resin. Tommy then weighs the residue (the remaining material). He compares the weight of the residue to the initial weight of the sample to determine the percentage of glass in the sample.

Tommy uses an instrument called an acoustic emission tester for the acoustical test. He connects the acoustic emission tester to a sample and applies a mechanical load to the sample. By noting the overall pattern of sounds emitted by the sample, Tommy can determine whether the sample has the mechanical properties required for application.

Purpose

In this lab, you will make and test a composite material.

Lab Objectives

When you've finished this lab, you will be able to—

- Demonstrate a method for making fiberglass.
- Compare the strength of two related materials.

Lab Skills

You will use these skills to complete this lab—

- Measure length and width using a ruler.
- Measure volume using a graduated cylinder.
- Test performance of a material.

Materials and Equipment Needed

fiberglass repair kit
paintbrush
wax paper
4-oz paper cups (unwaxed)
scissors
acetone
heavy object
gloves
goggles
marking crayon
string
weights
bench clamp
meterstick or yardstick
machinist's rule

Pre-Lab Discussion

During a chemical reaction, there may be a change in heat content; for example, the reaction either absorbs heat from the surroundings or releases heat to the surroundings. The amount of heat absorbed or released depends on the reactants and the products of the reaction. For instance, if sulfur is combined with oxygen to produce sulfur dioxide, heat is released. The equation for this reaction is written

$$\underset{\text{(sulfur)}}{S} + \underset{\text{(oxygen)}}{O_2} \longrightarrow \underset{\text{(sulfur dioxide)}}{SO_2} \qquad \Delta H = -70.96 \text{ Kcal/mole}$$

Equation L10-1

H stands for heat content and Δ is the math symbol for "change in." So the expression $\Delta H = -70.96$ Kcal/mole indicates that the change in heat content for the reaction is negative, i.e., heat is released to the surroundings.

Some chemical reactions have positive changes in heat content. For instance, the chemical reaction for the decomposition of potassium chloride is

$$2KCl \longrightarrow 2K + Cl_2 \qquad \Delta H = +\ 104.17\ \text{Kcal/mole}$$

(potassium chloride) (potassium) (chlorine gas)

Equation L10-2

The positive ΔH indicates that the decomposition of KCl absorbs heat from the surroundings. (To get a sample of KCl in a test tube to decompose, you must heat it.)

In this lab you will use a fiberglass repair kit containing polyester resin, a curing agent, and a glass-fiber mat. These materials are used by auto body technicians to repair damage to car bodies made of fiberglass.

To make fiberglass, you will first mix the resin with the curing agent. The chemical reaction that occurs will cause the polyester chains of the resin to become cross linked. When this happens, you will notice several changes in the properties of the resin. You should also notice an energy change that takes place during the reaction. This change relates to the heat content of the reactants and products in the reaction.

When the chemical reaction between the resin and the curing agent is complete and the resin hardens, you will apply the resin in layers to several pieces of glass fiber mat. The composite material you will have formed is called fiberglass. How will its strength compare to that of the resin or glass fiber alone? You will have the opportunity to test your answer to this question.

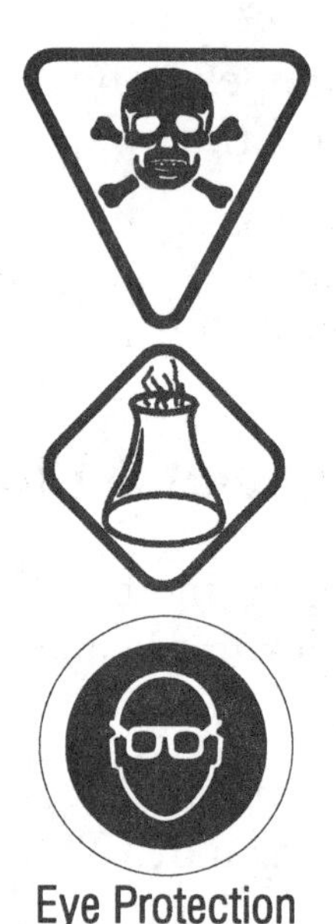

Safety Precautions

- Do not inhale polyester resin or curing agent or get it on your skin, in your mouth, or in your eyes.
- Work in a well-ventilated area while making fiberglass (preferably under a fume hood).
- Wear goggles when testing materials for strength.

LAB PROCEDURE

Method

Put on your lab apron and goggles.

Part A. Making a Fiberglass Composite

Work in a well-ventilated area and wear goggles and gloves.

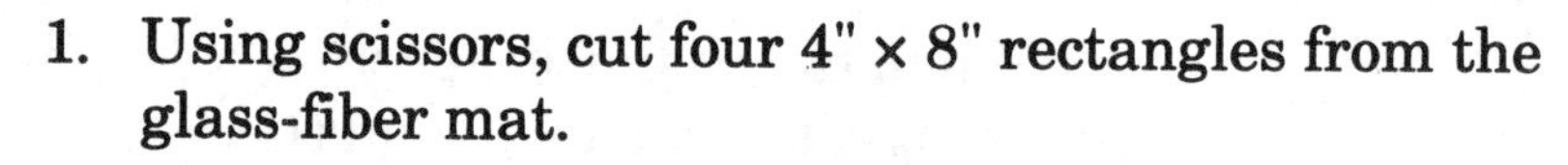

1. Using scissors, cut four 4" × 8" rectangles from the glass-fiber mat.
    - Place a piece of wax paper (12" × 16") on a clean, level counter.
    - Using a marking crayon, draw the outline of a 4" × 8" rectangle in the center of the wax paper.

2. Prepare the matrix of your composite by mixing the resin and curing agent according to the instructions in your kit. When you have finished mixing, you will have only about 5 minutes to work with the matrix before it sets.
3. When the resin is prepared, do the steps below, working quickly but carefully. You do not want the resin to set up before you have finished your composite.
    - Using the paintbrush, spread some resin evenly over the rectangle on your wax paper until it is well coated.

- Place one 4" × 8" piece of fiber mat directly over the spread resin.
- Dip brush into resin and transfer resin onto the center of the fiber mat. Brush out from the center of the mat until the entire mat is saturated with a heavy coat of resin.
- Add a second piece of fiber mat directly over the first, and coat with resin as above.
- Add and coat the remaining fiber mats in the same way.

4. When you have finished preparing the composite, place a 12" × 16" piece of wax paper firmly over it. (Do not wrinkle the wax paper.) Then place a heavy object over the slab so that the weight is evenly distributed over the surface. While you are waiting for the composite to cure, go to the next step.
5. Lay a clean piece of wax paper (12" × 16") on a clean, level counter. Then do the following:
   - Using a marking crayon, draw the outline of a 4" × 8" rectangle in the center of the wax paper.
   - Prepare more resin as you did in Step 2.
   - Before the matrix sets, pour it slowly onto the wax paper until you have evenly covered the rectangle. Work slowly to keep the resin inside the rectangle.
6. Wait until the surface of the slab has set, then gently place a 12" × 16" piece of wax paper over it. Place a heavy object onto the slab so that the weight is evenly distributed over the whole surface.
7. Allow the slabs to cure for a minimum of 2-3 hours.

Part B. Testing Properties of a Fiberglass Composite

Combine two lab stations to make a group of four students.

1. Lift the slabs of fiberglass and matrix from the wax paper.
2. Using the crayon, place 2"-long marks 1" from each end of both slabs (See Figure L10-1).

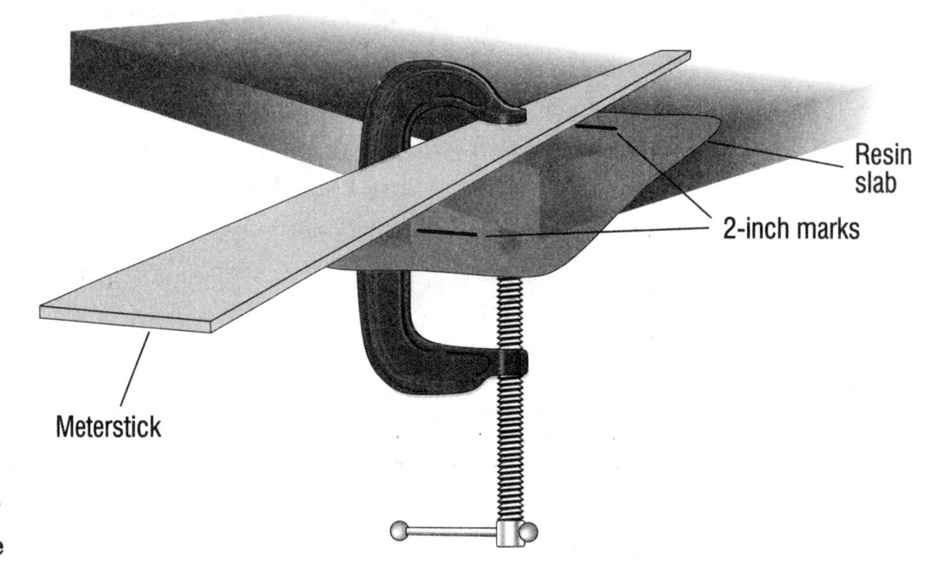

Figure L10-1 Testing apparatus for the composite

3. Set up the testing apparatus shown in Figure L10-1, securing the resin slab and meterstick firmly to the lab counter with the clamp. Be sure the meterstick is level. Then, using the machinist's rule, measure the initial gap between the meterstick and the slab as shown in Figure L10-2. Record this value in the data table.

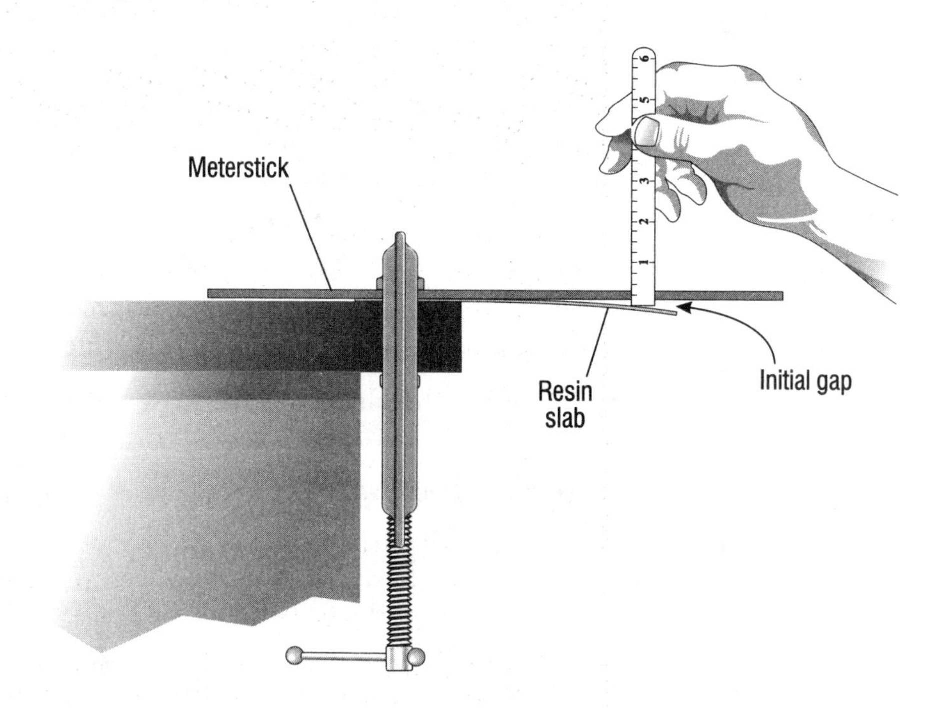

Figure L10-2 Measuring deflection of the slab

4. After you have put on your goggles, add weights to the string, 100 grams at a time, until you have added all the weight available. Then do the following:
   - Record the total weight applied in the data table.
   - Using the machinist's rule, measure the final gap between the resin slab and the meterstick, using the same technique as in Figure L10-2. Record this value in the data table.
5. Wearing your goggles, repeat steps 3 and 4 for the fiberglass slab. Record values in the data table and do the Calculation.

Cleanup Instructions

- Discard unwanted slabs of resin or fiberglass in the trash.
- Pour some acetone into a paper cup and clean your paintbrush. Leave under the hood to dry.
- Discard cups of unused resin in the trash.
- Return pieces of unused glass fiber mat to storage.

Observations and Data Collection

Data Table

Materials	Initial Gap	Weight Applied	Final Gap	Deflection
Resin slab				
Fiberglass slab				

Calculation

Calculate the deflections of the resin slab and fiberglass slab during testing using the following formula:

$$\text{Deflection} = \text{Final Gap} - \text{Initial Gap}$$

Record these deflections in the data table.

WRAP-UP

Conclusions

1. What was your first evidence that the resin underwent a chemical reaction after the curing agent was added? What other indications of a chemical reaction did you observe? Was there a change in heat content during the reaction? Explain.
2. Show with a drawing how the fibers of your composite are oriented inside the matrix. In your drawing, label the components of the composite with the names of the compounds you used. Also label the type(s) of composite you made (flake, particle, fiber, laminar).
3. For which material in Part B did you expect to measure more deflection under load? Why did you expect this? Did the test results agree with what you expected? Provide evidence for this from the data you collected. What can you conclude about the difference in properties of the resin slab and fiberglass composite?

Challenges

4. Design a way to determine whether the materials tested in Part B failed to return to their original form following the test. What do the results of this test indicate further about the difference in properties between the resin slab and composite?
5. Write a procedure for the use of the fiberglass repair kit for an auto body repair person to follow in repairing a hole in an auto door panel.

UNIT WRAP-UP ACTIVITY

Throughout this unit, you have completed activities that have had you examine various materials, their properties, and their uses. In this activity you will draw on your results from one or more of these activities: 1-1, 1-2, 1-3, 2-1, 2-5, 3-9, 4-6, 5-2, 5-4. You will work in small groups.

Your group has been assigned the task of selecting a product for reengineering using one or more of the synthetic materials discussed in this unit to improve performance and durability. Within your group, consider the following questions and record your answers:

- What items do you, your family, and your friends use frequently?
- What materials are currently used in the product?
- Which products could be improved by using different materials?

Choose one of the products and focus on the following questions:

- What characteristics of the product's performance and durability would you like to change?
- Should the material conduct heat? Should it retain heat or keep items cold?
- Should the material resist harsh chemicals?
- Should the material conduct electricity or act as an insulator?
- Should the material hold its shape or be flexible?
- How long should the material last?
- Is weight a factor?
- Which of the synthetic materials covered in this unit will best suit your needs?
- How will the change in materials affect the cost of the product?

When you have decided on a product to reengineer, prepare a presentation for the class. Include in your presentation a model that indicates the changes you have made or propose to make. Use the model to explain the improvements in performance and durability that will result from using the materials you have chosen. Tell the class of any test that can be used to test the desired properties of the materials you are using.

GLOSSARY

acid — a substance that increases the hydrogen ion concentration in a solution

acid-base neutralization — said of a chemical reaction in which the amounts of acid and base react to form water (and salt) and leave no excess of H_3O^+ or OH^- ions

alkali — refers to a family of metals that is located in group 1 of the periodical table. These metals are so named because of their oxides and hydroxides are strong bases.

alloy — a synthetic material produced when a metal is combined with another metal or a nonmetal

amorphous — relating to a solid that is noncrystalline with no definite form or structure

anion — any atom or group of atoms with a negative charge

anneal — the process of treating metals, alloys, or glasses with heat and then cooling to remove stress and make the material less brittle

anode — positive terminal of a primary cell or storage battery

antioxidant — an inhibitor that is effective in preventing oxidation by molecular oxygen

arthritis — an autoimmune disease causing inflammation or swelling of the joints

atom — the smallest particle of an element, made of a positively charged nucleus surrounded by negatively charged electrons

atomic diameter — diameter of an atom—calculations based on distances between nuclei in a variety of compounds containing the atom

atomic mass — the mass of one atom of a given element

atomic number — number of protons in the nucleus of an atom (equal to the number of electrons in a neutral atom)

atomic weight — the average mass of an atom of an element as it occurs in nature

base — a substance that, when mixed with water, forms hydroxide ions

base — one of the three subunits of a nucleotide found in DNA and RNA molecules

biocompatible — said of a substance that interacts with the body in a harmonious way

biomaterial — nonliving substance that is used to take the physical place of or replace the function of living tissues

blow molding — a method of producing a hollow object by inflating a blob of hot melt inside a mold

boiling point — the temperature at which the vapor pressure of a liquid exactly equals the pressure above the liquid; i.e., the point at which the liquid has entered the gas phase

bond — attractive force that holds atoms and ions together in molecules

cancer — a large group of diseases that involve the uncontrolled growth and spread of abnormal cells in the body

carcinogen — agent involved with the onset of cancer, such as chemicals or radiation

cardiovascular system — the system of heart and blood vessels that delivers blood to issues throughout the body

cartilage — tough, fibrous connective tissue attached to the sides of bone

cast — to form a liquid or plastic substance into a specific shape by letting it cool in a mold

cast iron — any carbon-iron alloy cast to shape and containing 1.8-4.5% carbon

cathode — positively charged pole of a primary cell or storage battery

cation — any atom or group of atoms with a positive charge

cellulose — a complex carbohydrate found in stems, bark, and fibrous parts of plants that is not digestible by animals

ceramic — product made by the baking or firing of nonmetallic elements

characteristic — a distinguishing trait, quality, or property

chemical equation — an expression that uses numbers and symbols to represent a chemical reaction; reactants are shown on the left while products are shown on the right.

chemical formula — a description of a molecule that shows the type and number of atoms it contains

chemical property — relates to its ability to enter into or resist a chemical change

chemical reaction — changing of substances to other substances by breaking existing chemical bonds and forming new chemical bonds

clotting — the change of blood from a liquid to a gelatin-like substance as a result of chemical reactions occurring in the blood

collagen — a protein that gives the body's connective tissues their structure

combination reaction — in this reaction two or more elements or compounds combine to form a new compound

composite — material composed of combinations of metals, alloys, ceramics, or polymers with one material serving as a matrix inside which another material is embedded

compound — two or more different elements joined together by chemical bonds

compression molding — making plastics in a mold using heat and pressure to form the material

compressive strength — ability of a stationary block to resits breaking by pushing forces

condensation reaction — a reaction that joins two molecules into one molecule with the elimination of water

conductivity — ability of a substance to transfer energy, usually in the form of heat or electrical energy

congenital — existing at or dating from birth

corrosion — the oxidation or destruction of metals

covalent bond — a chemical bond between atoms in which the electrons are shared

covalent bonding — the bonding of atoms in which each atom of the bond pair contributes one electron to form a pair of electrons

cross-link — the linking together of the parallel chains of a large molecule, such as a polymer

crystal — a three-dimensional atomic, ionic, or molecular structure consisting of repeated, identical unit cells

crystal lattice — the regular arrangement of atoms in a crystalline solid

crystalline — resembling or composed of crystals

decomposition — the breakdown of nonliving organic matter by organisms, especially microorganisms

deflocculent — an agent that removes floc, a mass formed by the clumping of particles

densification — the increase in the density of a material

density — a measure of the relationship between the mass and the volume of a substance; mass per unit volume of a substance

detritus — decaying organic material

devitrified — converted from a glassy texture to a crystalline texture

diamagnetic — having a magnetic permeability of less than 1; these materials are repelled by a magnet.

die — a tool or mold used to impart shapes to or to form impressions on materials such as metals and ceramics

dielectric — an electric insulator material

dielectric heating — heating in an insulator while opposing an alternating electrical current

diffusion — random movement of molecules due to the energy of the molecules themselves that cause solutes and solvents in solutions to move from areas of higher concentration to areas of lower concentration

distillation — a process of producing vapors from a liquid by heating the liquid in a vessel and then collecting and condensing the vapors into liquids

double displacement reaction — involves the exchange of positive and negative ions between two compounds

ductility — ability of a material to be drawn into a wire, hammered thin, or easily molded without breaking

elastomer — a material that can be stretched under low stress to at least twice its original length and that will return to its original form when released

electric current — net transfer of electric charge per unit time

electricity — a physical phenomenon involving electrical charges and their effects at rest and in motion

electrolyte — a substance that, when dissolved in water, conducts an electric current

electron — a negatively charged subatomic particle found outside the nucleus of an atom

electronegativity — the relative ability of an atom to attract an electron to itself

element — one of the basic substances that make up the world, determine the characteristics of all materials, cannot be separated into different substances by chemical processes, and can be changed to other elements only by radioactivity or nuclear reaction

embolism — the blocking of a blood vessel by an abnormal blood clot circulating in the blood

equilibrium — a state of balance between two opposing forces; a state where the rate of particles entering and leaving the system is equal

erosion — loss of soil from an area because of flooding, wind, or rainfall

evaporate — to change into a vapor

extrusion — process in which a semisoft solid material is forced through a die to produce a continuously formed piece of material in the shape of the desired product

fatigue strength — ability to resist breaking due to repeated loading and unloading

ferromagnetic — strongly attracted by a magentic fields

ferrous — composed of the ionic form Fe^{+2} of iron and of other elements; relating to iron

flammability — ability of a substance to ignite and burn quickly

forging — using compressive force to shape metal by plastic deformation

free radical — a highly reactive atom that has one or more unpaired electrons

gas chromatography — separation technique involving passage of a gaseous moving phase through a column containing a fixed absorbent phase

gravitational force — the force of attraction between two bodies because of their masses

high-performance liquid chromatography — separation of a mixture carried through a packed column by a solvent, based on adsorption of molecules to a solid packing material

homogeneity — the quality of having similar properties throughout the whole of an object

hypersensitivity — an extreme allergic reaction to foreign substances

inductive heating — heating in a conductor due to changes in electromagnetic induction while conducting alternating electrical current

inert gas — gas that is unable to react, such as helium or other gases in Group VIIIA of the periodic table

infrastructure — the underlying foundation or basic framework

ingot — a mass of metal cast into a shape that makes it easier to handle

injection molding — molding of metal, plastic, or ceramic shapes by injecting a measured quantity of the molten material into dies

inorganic — said of a chemical compound that does not contain carbon as a principal element

insulator — a material that is a poor conductor of heat, sound, or electricity

ion — a charged atom that may be either positive (because it has lost electrons) or negative (because it has gained electrons)

ionic bonding — a type of chemical bonding in which electrons are transferred completely from one atom to another

ionization potential — the energy per unit charge needed to remove an electron from a given kind of atom or molecule to an infinite distance

isotropic — describes a material with the same properties in all directions

kinetic energy — energy involved in producing work or motion

machining — performing various cutting or grinding operations on a piece of work

magnetic field — the field produced by a magnet that attracts and repels other magnetic materials

magnetic resonance imaging (MRI) — a technique that gives a computer image of body tissues by inducing their molecules to give off radio waves

malleability — capability of a material to be bent, shaped, or drawn out, as with hammering

matrix — a material in which something is embedded or enclosed

melting point — the temperature at which a solid melts to become a liquid

metal — an electropositive element that is usually ductile and malleable with high tensile strength

metallic bonding — chemical bond in metals, produced by sharing valence electrons between atoms in a stable crystalline structure

metallurgist — a scientist or engineer who studies metals and their properties

microstructure — the structure of an object as revealed by a microscope

modulus of elasticity — the ratio of the stress in a material to the strain that it produces

molecular formula — the number of each type of atom in a molecule of a compound

molecular structure — geometry and arrangement of atoms in a molecule

molecule — the smallest possible form of any compound, made of atoms bonded to each other

molten — made liquid by heating

monomer — a molecule that can be combined with another molecule to form a polymer

neutralize — to make a solution neutral in pH

neutron — an uncharged particle found in the nucleus of an atom

nonferrous — any metal other than iron and its alloys

nonmetal — an element that is usually negatively charged ions or radicals, capable of combining with oxygen to form oxides

nucleus — positively charged region of an atom made up of protons and neutrons and accounting for almost all of the mass of an atom

nucleus — the control center of a cell surrounded by its own membranes

orbital — the region of space around the nucleus of an atom that can be occupied by one or two electrons

orbital magnetic moment — the magnetic dipole moment associated with the motion of a charged particle about an origin

oxidation — process in which a substance combines with oxygen, such as combustion, rust, and the burning of calories in the body; the loss of electrons by a substance, typically when it combines with oxygen

oxidation-reduction reaction — a chemical reaction in which one reactant is oxidized and another reactant is reduced. Also called a redox reaction.

oxidizability — is a measure of a materials ability to readily lose one or more electrons

paramagnetic — describes a material that is repelled by both magnetic poles of a magnet and therefore aligns perpendicular to the magnetic lines of force

periodic table — table that contains all the known elements, arranged according to their chemical bonding characteristics

petrochemical — a chemical derived from fossil fuels

physical property — is a given characteristic that can be measured without a change in the chemical makeup of the material that is tested

physiology — study of life processes and functions of organisms

piezoelectric — having the ability to generate a voltage when a mechanical force is applied or to produce a mechanical force when a voltage is applied

platelet — a fragment of a blood cell membrane that functions in blood clotting

polar — said of a molecule having a separation of charges resulting in a positive and a negative region of the molecule

polarity — is the separation of charges in a molecule

polarize — to separate a positive and negative electrical charge and create an electric dipole

polarization — the action of polarizing or state of being or becoming polarized

polymer — a large molecule consisting of many identical or similar monomers linked together

polymerization — a chemical reaction in which two or more simple molecules combine to form a more complex molecule made up of repeating units of the simple molecules

porosity — a measure of how porous a material is

product — a compound that is the result of a chemical reaction

property — are characteristics of a given material

proton — a positively charged particle in the nucleus of an atom

pyrolysis — a heating process—in a sealed reactor without oxygen—that breaks down synthetic rubber into a form of heating oil

quenching — cooling by immersing in water or oil

radiation — a form of energy that includes visible light, infrared light, ultraviolet light, X rays, and gamma rays; the means by which body heat is lost in the form of infrared rays

raw material — a crude, unprocessed or partially processed material used to manufacture a product

reactant — substance that enters into and changes during a chemical reaction to produce products

reactivity — ability to participate in a chemical reaction

reactor — vessel in which a chemical reaction takes place

redox — relating to oxidation-reduction reactions

reducibility — is a measure of a materials ability to readily gain one or more electrons

reduction — the gain of electrons by a substance during a chemical reaction

refractory — resisting heat

resistivity — the resistance of a material to the flow of electrical current times the cross-section of the material per unit length of material

rotation — movement about an axis or center

saturated — said of a solution that is as concentrated as possible

saturated — said of fats or fatty acids that have no carbon-carbon double bonds

semiconductor — a solid whose electrical conductivity is between that of a conductor and that of an insulator

shear strength — a material's ability to resist forces that cause parts of the material to slide past one another

shock loading — ability to resist breaking due to a sudden change in loading

single displacement — one element replacing another element

single displacement reaction — involves one elements replacing another in a compound producing a new compound and a different element

sinter — to heat (but not melt) a powder to form a coherent mass

sol-gel process — a solution-based technique by which ceramics can be made at low temperatures with short firing times

solution — mixture in which one substance (the solute) is dissolved and the other substance is the dissolving medium (the solvent)

solvent — a substance that can dissolve another substance

specific heat — the amount of energy required to raise the temperature of one gram of a substance by one Celsius degree at 15°C

specific heat capacity — a measure of the calories per unit mass of substance required to raise its temperature by one unit of temperature measure

spin-magnetic movement — magnetic moment due to the spin of an electron

sterilization — a surgical procedure that renders an organism incapable of producing or releasing viable sperm or ova

sterilization — killing all microorganisms including viruses and spores

superconductor — a substance that shows no resistance to electrical current at extremely cold temperatures (temperatures near 0 K)

synthetic — not naturally occurring; manmade

tensile strength — strength of a material against a pulling force

thermal — of or relating to heat

thermal conductivity — the ability to conduct heat

thermal expansion — the amount of change in volume that results from a change in temperature

thermoplastic — capable of softening when heated and then hardening when cooled; said of a polymer that melts when heated

thermosetting — able to become permanently solid when heated; said of a polymer that chemically changes during processing and becomes permanently solid

transfer molding — is a modification of compression molding. It eliminates the turbulence and uneven polymer blow caused by compression molding.

transition elements — are elements located in the periodic table between groups 2 and 13 and are in periods 4 and greater

transition metals — same as transition elements

translation — the process of assembling amino acids into proteins at the ribosome according to the instructions carried by messenger RNA

translation — uniform motion of a body in a straight line

valence electrons — electrons in the outermost orbitals of an atom

van der Waals attraction — the attractive force between atoms or between nonpolar molecules due to induced charge separation

vaporize — to change from a liquid to a vapor

vibration — the continuing periodic change in the positions of atoms and in molecules and/or crystals relative to each other

vibration characteristics — amplitude (or how large the vibration is) and frequency (how fast the vibration occurs) are traits that describe vibrations

viscosity — resistance to flow in a gas or a liquid

vulcanization — process of treating rubber or plastic materials to give them useful properties

weight — a measure of the force that pulls an object of a certain mass toward the center of Earth

welding — uniting metallic parts by heating and allowing the metals to flow together

Notes: